CAMBRIDGE LIBRARY COLLECTION

Books of enduring scholarly value

Technology

The focus of this series is engineering, broadly construed. It covers
technological innovation from a range of periods and cultures, but centres on
the technological achievements of the industrial era in the West, particularly
in the nineteenth century, as understood by their contemporaries.
Infrastructure is one major focus, covering the building of railways and
canals, bridges and tunnels, land drainage, the laying of submarine cables,
and the construction of docks and lighthouses. Other key topics include
developments in industrial and manufacturing fields such as mining
technology, the production of iron and steel, the use of steam power, and
chemical processes such as photography and textile dyes.

Life of Richard Trevithick

Cornishman Richard Trevithick (1771–1833) was one of the pioneering
engineers of the Industrial Revolution. Best remembered today for his
early railway locomotive, Trevithick worked on a wide range of projects,
including mines, mills, dredging machinery, a tunnel under the Thames,
military engineering, and prospecting in South America. However, his
difficult personality and financial failures caused him to be overshadowed
by contemporaries such as Robert Stephenson and James Watt. This
two-volume study by his son Francis, chief engineer with the London and
North-Western Railway, was published in 1872, and helped to revive his
neglected reputation. It places its subject in his historical and technical
context, building on the work of his Father, Richard Trevithick Senior, and
the Cornish mining industry. It contains much technical detail, but is still
of interest to the general reader. Volume 2 continues examining his work
thematically, and includes his work in Peruvian mines.

Cambridge University Press has long been a pioneer in the reissuing of out-of-print titles from its own backlist, producing digital reprints of books that are still sought after by scholars and students but could not be reprinted economically using traditional technology. The Cambridge Library Collection extends this activity to a wider range of books which are still of importance to researchers and professionals, either for the source material they contain, or as landmarks in the history of their academic discipline.

Drawing from the world-renowned collections in the Cambridge University Library, and guided by the advice of experts in each subject area, Cambridge University Press is using state-of-the-art scanning machines in its own Printing House to capture the content of each book selected for inclusion. The files are processed to give a consistently clear, crisp image, and the books finished to the high quality standard for which the Press is recognised around the world. The latest print-on-demand technology ensures that the books will remain available indefinitely, and that orders for single or multiple copies can quickly be supplied.

The Cambridge Library Collection will bring back to life books of enduring scholarly value (including out-of-copyright works originally issued by other publishers) across a wide range of disciplines in the humanities and social sciences and in science and technology.

Life of
Richard Trevithick

With an Account of his Inventions

VOLUME 2

FRANCIS TREVITHICK

CAMBRIDGE UNIVERSITY PRESS

Cambridge, New York, Melbourne, Madrid, Cape Town, Singapore,
São Paolo, Delhi, Dubai, Tokyo, Mexico City

Published in the United States of America by Cambridge University Press, New York

www.cambridge.org
Information on this title: www.cambridge.org/9781108026680

This edition first published 1872
This digitally printed version 2011

ISBN 978-1-108-02668-0 Paperback

LIFE

OF

RICHARD TREVITHICK,

WITH AN ACCOUNT OF HIS INVENTIONS.

By FRANCIS TREVITHICK, C.E.

ILLUSTRATED WITH ENGRAVINGS ON WOOD BY W. J. WELCH.

VOLUME II.

LONDON:

E. & F. N. SPON, 48, CHARING CROSS.

NEW YORK:

446, BROOME STREET.

1872.

LONDON: PRINTED BY WILLIAM CLOWES AND SONS, DUKE STREET,
STAMFORD STREET AND CHARING CROSS.

CONTENTS OF VOLUME II.

CHAPTER XVII.

VARIOUS INVENTIONS.

CHAPTER XVIII.

AGRICULTURAL ENGINES; LOSS OF PAPERS.

CHAPTER XIX.

Pole Steam-engine.

CHAPTER XX.

The Watt and the Trevithick Engines at Dolcoath.

CHAPTER XXI.

ENGINES FOR SOUTH AMERICA.

CHAPTER XXII.

PERU.

CHAPTER XXIII.

Costa Rica.

CHAPTER XXVI.

TUBULAR BOILER — SUPERHEATING STEAM — SURFACE CONDENSER.

CHAPTER XXVII.

HEATING APPARATUS — MARINE STEAM-ENGINES — REFORM COLUMN.

ILLUSTRATIONS TO VOLUME II.

LIFE OF TREVITHICK.

CHAPTER XVII.

VARIOUS INVENTIONS.

"About 1804 Captain Trevithick put up in Dolcoath Mine a stone-crushing mill, having large cast-iron rollers, for breaking into small pieces the large stones of ore; it was spoken of as the first ever used for such a purpose; the same form of crusher is still used in the mines. It caused a great saving compared with breaking by a hand hammer."[1]

"I saw at the Weith Mine in 1805 a portable high-pressure engine, made by Captain Trevithick.

"It was called a puffer; the cylinder was in the boiler; the steam about 30 lbs. on the inch above the atmosphere. A wooden shed sheltered the engine and man.

"The facility of manufacture and cheapness of those engines caused them to be much used in the mines, and also elsewhere."[2]

Mrs. Trevithick, about the time we are speaking of, accompanied her husband through one of the Staffordshire china manufactories. Trevithick said to the manufacturer, "You would grind your clay much better by using my cast-iron rolls and high-pressure steam-engine." The manufacturer begged him to accept a set of china. Mrs. Trevithick was disappointed at hearing her husband say "No! I have only told you what was passing in my mind."

Driving rolling-mills was among the early applications of the high-pressure steam-engines; but pulverizing

---

[1] Recollections of the late Captain Charles Thomas, manager of Dolcoath.
[2] Captain Samuel Grose's recollections.

hard rock by the use of iron rollers was a novelty :
though his patent of 1802 shows the proposed rolls
driven by steam for crushing sugar-canes, yet no one
had dreamt, prior to 1804, of economy in crushing
stone and clay by such a means. The plan, however,
remains in use to this day in many mines, and is frequently spoken of under the name of quartz-crusher.

"MR. GIDDY,   "COALBROOKDALE, *September* 23rd, 1804.

"Sir,—Yours of the 13th this day came to hand. I left
Wales about eight weeks since, and put an engine to work in
Worcester, of 10-horse power, for driving a pair of grist-stones,
and a leather-dressing machine, and another in Staffordshire
for winding coals; each of them works exceedingly well.

"From Coalbrookdale I went to Liverpool, where a founder
had made two of them, which also worked exceedingly well;
one other was nearly finished, and three others begun. Some
Spanish merchants there saw one of them at work, and said
that as soon as they returned to Spain they would send an order
for twelve engines, of 12-horse power, for South America. In
South America and the Spanish West Indies water is very
scarce; in several places there is scarcely water for the inhabitants to drink, therefore there is no water for any engine. By
making inquiry, I found that ten mules would roll as much cane
in an hour as would produce 250 gallons of cane-juice, which
they boil until the water is evaporated, and the sugar produced.

"I told them that the engine-boiler might be fed with this
juice, and by a cock in the bottom of the boiler constantly
turning, and by taking a greater or smaller stream from it,
they might make the juice as rich as they liked. In this process the juice would be so far on towards sugar, and the fire
that worked the engine would cost nothing, because it would
have taken the same quantity of fuel under the sugar-pans to
evaporate the water, as it would in the engine-boiler.

"The steam from the engine might be turned around the
outside of the furnace for distilling rum, as the distilleries
require but a slow heat.

" I think the steam would answer a good purpose around the outside of the pan.

"If this method answers, the cost of working the engine would be nothing, and the engine would be then working, as it were, without fire or water.

"The Spaniards told me that if this plan answers, they would take a thousand engines for South America and the Spanish West Indies. I shall be very much obliged to you for your opinion on this business. These merchants make a trade of buying up sugar mills and pans, with every other thing they want from England, and exchange them with the Spaniards for sugar.

" At Manchester I found two engines had been made and put to work; they worked very well: three more are in building. From there I went to Derbyshire. The great pressure-engine I expect will be at work before the middle of October. A foundry at Chesterfield is building a steam-engine as a sample; two foundries in Manchester are at full work on them, and one in Liverpool. There are six engines nearly finished at Coalbrookdale, and seven in a foundry at Bridgenorth.

"I am making drawings for several other foundries. Any number of them would sell. A vast number are now being erected, and no other engine is erected where these are known. The engine for the West India Docks was neglected during my absence from the Dale, but I expect it will be ready to send off in ten days.

" In about three weeks I shall be in London to set it up. It will please you very much, for it is a very neat and complete job, and I have no doubt will answer every purpose exceedingly well. At Newcastle I found four engines at work, and four more nearly ready; six of these were for winding coal, one for lifting water, and one for grinding corn.

"That grinding corn was an 11-inch cylinder, driving two pair of 5-feet stones 120 rounds per minute; ground 150 winchesters of wheat in twelve hours with 12 cwt. of small coal. It worked exceedingly well, and was a very complete engine, only the stroke was much too short, not more than 2 feet 6 inches, which made very much against the duty.

" The other engine that was lifting water had a 5½-inch

diameter cylinder, with a 3-feet stroke, drawing 100-gallon barrels, twenty-four every hour, 80 yards, burning 5 cwt. of coal in twenty-four hours.

"This work it did with very great ease. I believe you will find this an exceeding good duty for a 5½-inch cylinder engine.

"Below I send a copy of Mr. Homfray's and Mr. Wood's letters to me :—

"Mr. Homfray's, of the 10th September.—' Our great engine goes on extremely well here, nothing can go better ; the piston gives no trouble ; it goes about three weeks, and we work it with blacklead and water ; the cylinder is as bright as a looking-glass ; it uses about 2 lbs. of blacklead in a week ; about once in twelve or fifteen hours we put a small quantity of blacklead, mixed with a little water, through the hole in the cylinder screw, and we never use any grease. We rolled last week 140 tons of iron with it, and it will roll as fast with the both pair of rolls, as they can bring to it.'

"Mr. Wood's letter, September 12th.—' We are going on, as it is likely we always shall, in the old dog-trot way, puddling and rolling from the beginning of the week till the end of it. Your engine is the favourite engine with every man about the place, and Mr. Homfray says it is the best in the kingdom.'

"I have not the smallest doubt but that I can make a piston without any friction or any packing whatever, that needs not to have the cylinder screw taken up once in seven years. It is a very simple plan, and will be perfectly tight; it is by restoring an equilibrium on both sides of the piston. I expect to see you in London soon, and then will give you the plan for inspection before I put it in practice.

"I am very much obliged to you for recommending these engines in Cornwall, but you have not stated in what manner they are to be applied; whether to work pumps, or barrels, or both. They may be made both winding and pumping engines at the same time, if so required.

"A rotative engine will cost more than an up-and-down-stroke, on account of the expense of the fly-wheel and axle. An engine capable of lifting 180 gallons of water per minute 20 fathoms would cost, when complete and at work, patent right included, about 220*l*. If it is a rotative engine, with a winding barrel, it will cost 270*l*. I expect that a 7-inch cylinder would be sufficient for winding at Penberthy Crofts, which might have a crank on the fly for lifting water in pumps, and a winding barrel on its back. This would cost about 170*l*.; the erection of them, when on the spot, will cost nothing. You do not say when you intend to be in town. I hope you will be present when the dock-engine is set to work.

"The engines first sent to Cornwall, must be from Coalbrookdale; then they will be well executed, but from Wales it would not be so.

"You may depend on having a real good engine sent down, with sufficient openings given to the passages.

"The engineer from the Dale has been lately in London, and has just returned; he gives a wonderful account of the engines working in London. There are twelve now at work there. They have well established their utility in different parts of the kingdom, and any number would sell. The founders intend to make a great number, of different sizes, and send them to different markets for sale, completely finished, as they stand.

"You do not say anything about wheels to the engine for Penberthy Crofts. There are several engines here nearly finished; if they suit in size for Penberthy, one may be sent down in four or five weeks, otherwise it may be two months.

"I am, Sir,

"Your very humble servant,

"RICHARD TREVITHICK.

"Direct for me at the Talbot Inn, Coalbrookdale."

Trevithick worked hard and successfully in making his steam-engines useful, and firmly believed that he could and would make them universal labourers. Even the Spanish merchants, unacquainted with steam, talked of giving an order for several engines for South America; and their glowing account of the wide field open to him may have been instrumental to his going to

that country by making his engines known there. His
proposal to make the sugar-cane convert itself into sugar
by the use of his patent high-pressure steam-engine may
be more theoretical than practical; but many more
unlikely things have come to pass.

At that time several of his engines were at work in
Wales, Worcester, Staffordshire, Coalbrookdale, Manchester, Derbyshire, Liverpool, Cornwall, and New-
castle-upon-Tyne. Twelve were at work in London,
and so familiar were people with them, that founders
intended to construct them of different sizes, and send
them for sale at the large market or county towns;
their cost complete, ready for work, to be 200l., more
or less, according to size, with a range of application
unlimited. His one letter, casually written sixty-seven
years ago, mentions them as grinding corn, dressing
leather, winding coal, crushing sugar-cane, prepared to
boil sugar, and distil rum; pumping water, rolling
iron, railway locomotion, portable steam fire-engine,
portable steam-crane, mine engines on wheels; so that
it may almost be said he was not too sanguine in
hoping to send in 1804 a thousand of his engines to
South America, for in those cursory remarks he draws
attention to no less than thirty-six high-pressure steam-
puffers at work.

The Penberthy Croft Mine portable engine could be
placed on wheels or otherwise, according to the wish of
the purchaser, as though steam locomotion was an
every day occurrence in 1804.

"MR. GIDDY,   "CAMBORNE, *January* 13*th*, 1811.

"Sir,—From calculating the quantity of blast given to
a blast-furnace, I find a considerable quantity more of coal consumed by the same quantity of air in this way, than by the

usual way in common engine chimneys. Of course the more cold air admitted to pass through the fire, the more heat carried to the top of the stack. Crenver 63-inch cylinder, double-power, 8-feet stroke, with but one boiler, works five strokes per minute. This gives about 1600 square feet of steam per minute, and burns about 8 tons of coals in twenty-four hours. The stack for this boiler is 3½ feet square, and the draught rises 10 feet per second, and will set white paper in a flame at the top of it in about a minute. Therefore, this chimney delivers 7200 square feet of air per minute, which is four and a half times the quantity of heated air, at nearly four times the temperature of heat that there is of steam produced from the same fire, and delivered to the cylinder.

"A blast-furnace that burns 100 tons of coal per week is blown by a 5-feet diameter air-cylinder, 4-feet stroke, ten strokes per minute, double power, giving about 1600 square feet of air per minute, to consume 100 tons of coal, besides giving a melting heat to 350 tons of ore and limestone.

" Crenver engine has 7200 square feet of air to burn 56 tons of coal per week, which is above eight times the quantity of air used by air fire-places to what is used in a blast-furnace, and of course must carry off a great proportion of the heat to the top of the stack, that might be saved if the engine-fire was a blast instead of an air fire.

"But suppose the idea to be carried still further, by making an apparatus to condense and take the whole of the heat into the cylinder instead of its passing up the chimney. By having a very small boiler, and a blast-cylinder to blow the whole of the blast into the bottom of the boiler, under a cylinder full of small holes under the water, to make the heated air give all its heat to the water.

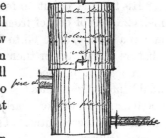

" The furnace must be made in a tight cast-iron cylinder. Both the fire-door and the hole through which the blast enters must be quite tight, as the pressure will be as strong in the fire-place as in the boiler. The whole of the air driven into the fire-place, with all the steam raised by its passage up through the water in the boiler, must go into the cylinder. There will also be the advantage of the expansion of the air by the heat over and above what it was when taken cold into the blast-cylinder.

"From the great quantity of coal burnt in blast-furnaces you will find that a very small blast-cylinder would work a 63-inch cylinder double. If there is as much heat in a square foot of air as in steam of the same temperature, the saving will be beyond all conception; but for my own part I cannot calculate from theory what the advantages will be, if any, and for that reason, before I drop or condemn the idea, I must request you will have the goodness, when you have an hour to spare, to turn your thoughts to this subject, and inform me of your sentiments on it.

"Perhaps it is like many other wild fancies that fly through the brain, but I did not like to let it go unnoticed without first getting your opinion. I hope you will excuse me for so often troubling you.

"St. Ives plans will be delivered to them on Tuesday, when I expect they will be forwarded to you.

"I hear there is a good course of ore in the adit end at Wheat Providence Mine.

"A Mr. Sheffield, of Cumberland, writes to Mr. Gould that he has turned idle his air-furnaces, and smelted his ores by a blast near a year since.

"His furnace is but 10 feet high and 4 feet diameter, and it melts 28 tons of ore, of from 4 to 5 in the 100 per week, and makes a regel of from 65 to 70 in the 100, and answers beyond what we calculated for them.

"Suppose a furnace 20 feet high and 4 feet diameter, it would smelt eight times the quantity of his, which would be near 900 tons per month, or nearly double the quantity raised by any one mine in the country. The expense of the . . . would be very trifling.

"To-morrow Dolcoath account will be held, when I expect to have orders to begin to erect a furnace on the spot.

"This trial of Mr. Sheffield's has put it out of my power to get a patent, and now I do not know how to get paid.

"I should be content with 5 per cent. on the profits gained by this plan, and would conduct the business for the mines without salary. Should you chance to fall on the subject with his Lordship, be pleased to mention something about the mode of my payment, as his Lordship is by far the properest person to begin with about my pay, for after his Lordship has agreed to the sum, and Dolcoath Mine the first to try the experiment, I think all the county will give way to what he might propose. But I wish something to be fixed on before all the agents in the mines know how to be smelters themselves, after which I expect no favour, unless first arranged.

"I remain, Sir,

"Your very humble servant,

"RICHARD TREVITHICK."

How great was the practical insight his genius gave him, and how imperfectly his followers have acted on this advice given sixty years ago!

The chimney that at its top would ignite paper, threw to waste four and a half times more heated air than was requisite to supply the quantity of heat which passed through the working cylinder in steam, and at a temperature nearly four times greater than the temperature of the steam. It needs only to observe the burnt appearance of a steamboat funnel of the present day to know how wasteful we still are, or how very ignorant of improved methods of economizing fuel.

To prevent this waste of heat up the chimney he proposed to do away with the chimney altogether; the fire-place was to be a close one, having a blast under the fire-bars of a. strength sufficient to force the air, heated by its passage through the fire, direct through a small valve into the water in the boiler, by which means all the heat given by ignition would pass into the steam, and his steam-puffer become an aërated steam-engine.

From the following it appears that this plan of Trevithick's is now coming into use as something quite new :—

" In your last impression, under the head of 'Air and Steam combined, as a Motive Power,' you state ' the invention was described to be that of Mr. Warsop, but we have recently heard that a few years back (1865) the same invention had been protected in an earlier patent than Mr. Warsop's, by Mr. Bell Galloway.'"[1]

Trevithick thought of patenting a plan for reducing copper ore by the use of a blast, in preference to the usual air-furnace and chimney, but something similar had been tried by Mr. Gould, and he therefore proposed to erect a blast-furnace in Dolcoath Mine, receiving a portion of the saving of fuel as his remuneration. Such a furnace worked there for many years, until copper smelting was removed from Cornwall to Wales. The plans for a breakwater at St. Ives were for an undertaking that has since been in many hands, but without success, except perhaps for the convenient making of members of Parliament. Some slight progress has been made by engineers and contractors, but vessels are not willingly taken to the port, and ratepayers grumble at unprofitable harbour taxes.

" DEAR TREVITHICK,   " LONDON, *January 20th*, 1811.

" I have not lost any time in mentioning your wishes respecting a compensation for the plan of smelting copper to Lord Dedunstanville, who intends mentioning the affair in his next letter to Mr. Reynolds. Lord Dedunstanville wishes you

---

[1] See the ' Mechanics' Magazine,' June 3rd, 1870.

extremely well, but it is impossible for him to settle anything apart from the adventurers.

"I am very sorry that anyone should have executed the plan of reducing copper ore by a blast-furnace before you had put into practice the idea suggested to me ten years ago. It ascertains, however, that the contrivance will succeed, although you are certainly reduced to ask moderate terms, and I know not what can be more moderate than those you have asked, except that I would recommend some limit as to time.

"The plan you suggest for an engine on a new construction is, I fear, very doubtful.

"According to the data furnished to me, the air in the blast would be to that in a common fire-place as $6\frac{1}{4}$ to 1 very nearly, provided their densities were the same; but you have measured one entering the furnace at the common temperature, and the other going to the stack so hot as to set on fire a piece of paper held at the top. Thus the increase of temperature that augments the elasticity of a fluid confined, would expand it in the same degree. It is therefore uncertain from these statements which furnace consumes the greater quantity of air. I apprehend the general principles of an engine worked by hot air, through the medium of a blast, would be as follows :—

"Let any quantity of air be driven into a furnace with the pressure of an atmosphere, and let it be there expanded ten times. It should then be taken off ten times as quick, but in that case no power whatever would be produced, so the external atmosphere would balance the internal. Now, let the blast be two atmospheres strong, and let them be expanded ten times, and be taken off ten times as fast, each stroke will be opposed by one, equal in all to ten; subtract two for the blast, there remain eight.

"But air so hot would burn every vegetable or animal substance, and such a furnace I suppose could scarcely be kept airtight. If the heated air is made to act on water, then it becomes a mere question of how much absolute heat is given out by the fuel, and whether that excess is more than sufficient to compensate the burden of the blast; for the water will absorb an immense quantity of heat in changing itself into steam, and thus reduce the force of the air as to make it almost impossible for that addition to add so much power as the blast takes away.

"I have, therefore, no hesitation in saying that this plan will certainly not do. Write to me by all means whenever anything strikes you, and you may always depend on having my best advice.

"I am, dear Sir,

"Ever most truly yours,

"DAVIES GIDDY."

Trevithick saw without apparent reasoning, while his friend's reasonings failed to make plain the full bearing of the questions, and so cramped the position as to make a change of front difficult—an operation in which Trevithick excelled. We learn, however, that in 1801 he suggested a blast in copper-ore furnaces, and in 1811 was on the verge of a discovery that has since revolutionized the iron-smelter's art by the use of hot blast. Wasted heat from a blast-furnace 10 feet high led him to the conclusion that by doubling the height of the furnace, enabling the cold mineral thrown in at the top to take up the heat wasted through the top of the low furnace, seven-eighths of the coal would be saved. His idea of sending blast through the furnace of his steam-boiler to economize heat could have been readily applied to the iron furnace, and we should have had the modern hot-blast iron furnaces.

[Rough draft.]

"GENTLEMEN,   "CAMBORNE, *March 5th*, 1812.

"Your favour of the 15th February, with a sketch of your brewery, I have received; and from which I find the head of water is 30 feet above the brewery, which makes it difficult to erect the chain and buckets so as to take advantage of the whole height of water; and as the stream is so very small, it will not admit of losing any part of the power.

"To erect a machine so high, to engage the whole fall, would be, I fear, more expense than the power you would get would warrant; therefore I would recommend it to be made use of in a cylinder, in the same way as we use falls of water of 200 feet in our Cornish copper mines. We allow one-third loss for friction and leakage in those machines; but your machine being so very small, the loss will exceed that proportion; therefore I cannot promise you above one-half of the real weight and fall to be performed on your machinery, and that must be by a well-executed machine, for a small defect would destroy the value of so trifling a power.

"As there is no expansion in water, it will be somewhat difficult to make the machine turn the centres with a fly-wheel, for if the valve shuts a little too early or too late for the turn of the crank over the centre, the fly-wheel's velocity must break something by confining the water between the piston and the bottom of the cylinder, which, after the valve is shut, cannot make its escape, and not having an elastic principle, the piston will strike as dead on the water as on a piece of iron, because, unless the valve is shut by the engine before the stroke is finished, it cannot shut at all.

"I know persons who have attempted to put fly-wheels on pressure-engines of this kind, but never yet has one been made to work rotative. I do not see much difficulty in making an engine of this kind to work a crank and fly-wheel, by connecting an air-vessel with the cylinder to receive the pressure and contract and expand and shut the valves, the same as in steam-engines.

"A machine on this plan ought to be placed as near the low level as possible. If I furnish you with drawings and directions for the executing of the work yourselves, I shall charge you fifteen guineas for them. If I send the machine finished, the charge will be 50l.

"Your objections respecting steam-engines I do not doubt are correct, when executed by persons who do not understand the construction of them. In England some persons privately erected my engine to evade the patent premium, but have severely paid for their saving knowledge by accidents and defects in their engineering ability. I have erected above 100 steam-engines on this principle, but never met with one accident or complaint against them. To prevent mischief from bad castings, or from the fire injuring the surface of cast iron, I make the boilers of wrought iron, and always prove them with a pressure of water, forced in equal to four times the strength of steam intended to be worked with.

"Some persons have worked those engines under a pressure above 100 lbs. on the square inch, but in general practice I do not exceed 20 lbs., finding under this pressure the piston will stand six or eight weeks, and the joints remain perfect, and no risk of bursting the boiler, it being made of wrought iron, and proved by pressure before sent off; but cast-iron boilers may, by defects not discernible, and are very apt to break by the water being left low in the boiler, and if heated red hot, exploding without the smallest notice; but wrought-iron boilers, when defective, give way only partially, without injury to anyone. With respect to the erecting and management of the engine, you need not have an engineer, for any common tradesman can do this from the drawings and directions sent with the engine; for, as I before informed you, farmers and their labourers set up and keep in order the thrashing-machine engines without my going on the spot or sending any person to assist them. I never saw a steam-engine rolling malt, therefore cannot judge the quantity the engine would roll, only by a comparison with horse labour, against the consumption of coal, which will be in some cases as about 42 lbs. to one horse; but where great speed is required in the machine, the coal will be less, as steam-engines make more revolutions in a minute than horse mills, therefore the work is done with less friction.

"I have several times applied the steam, after it has worked the engine, to boil water and other purposes, with as good effect as if the engine had not been there, therefore the work of the engine will be a clear profit.

"You say about a 1-horse engine. The boiler would be so small that it would not be worth applying that steam to any other purpose, as any large quantity of water would be but slowly heated.

"I find that it does not answer either the purpose of the vendor or the user of an engine, to make less than a 2-horse power, as the expense on a very small engine is nearly as much as one of the power I use for thrashing, those being only 80*l.*, and a 2-horse is 60*l.* Respecting the mashing with steam, I never before heard of it, but from the theory of the plan I think it cannot fail to answer a far better purpose than any other that can possibly be applied for extracting the essence of the malt. However, should it not answer your purpose, it is only the loss of the expense of a few yards of 1-inch lead pipe.

"In an engine of the size used for thrashing, if the fire is kept brisk, it will boil, by the steam sent into a separate vessel, near 300 gallons of water per hour.

"The room required to work in is about 7 feet diameter, and 12 feet high. It would be useless to put you to the expense of drawings, until you have made up your minds on what you intend to have done.

      "I remain, Gentlemen,

           "Your most obedient humble servant,

                "RICHARD TREVITHICK.

"To ROBINSON AND BUCHANAN, Brewers,
    "*Londonderry, Ireland.*"

Engineers of the present day do not volunteer such general information without charge, or give such a variety of practical mechanism slightly but clearly described, and principles reduced to practice. An endless chain with buckets is a form of water-wheel not then in use. A water-pressure engine for so small a quantity of water, with a fall of about 30 feet, would cause a loss of 50 per cent. from friction and small defects. The non-elastic character of water made it unsuitable for a machine requiring a fly-wheel. Air-vessels should be used to lessen the rigidity of water.   Cast-iron boilers dangerous.   Wrought-iron boilers to be tested with a water pressure four times as great as the proposed working steam pressure.   A steam pressure of 20 lbs. to the inch most suitable for engines in charge of inexperienced persons.   The brewers' mash tub to be heated by the waste or surplus steam.

[Rough draft.]

" Gentlemen,   " Camborne, *December 5th*, 1812.

"I have yours of the 20th November.   The letter you directed for Truro never came to hand.   I find by your letter that you have been trying to put into practice the hints I gave you about the chain and buckets, and that you expect it will answer if properly executed.   You are not the first that has picked up my hints, and stuck fast in their execution.   I make it a rule never to send a drawing until I have received my fee, and when you remit to me fifteen guineas I will furnish you with proper drawings and directions to enable you to make and erect the machine.

" I remain, Gentlemen,

" Your very humble servant,

" Richard Trevithick.

" Robinson and Buchanan, Brewers,
     "*Londonderry.*"

What a pocket encyclopædia of inventions! from which, as by stealth, Robinson and Buchanan selected the least applicable, declining a suitable steam-engine at a very small cost, rather than pay an engineer for his opinion.

[Rough draft.]

" Camborne, *April 26th*, 1812.
" Sir Charles Hawkins, Bart.,

" Sir,—I have received yours of the 7th, respecting the small breakwater at St. Ives.   As far as I can judge from a rough calculation, I think it an undertaking likely to pay well ;

but as you wished me not to mention anything about your intentions, and not receiving your orders to make a minute inquiry and estimate, I cannot answer your letter so fully as I should wish, fearing that giving a random and imperfect statement might be apt to lead you into errors, and also make me look simple. If an engineer were employed to survey and estimate after me, every information in my power is at your service; therefore be pleased to state particularly what information you wish, and I will attend to the business and answer your questions as early as possible.

"I have received a letter from Sir John Sinclair requesting correct drawings and statements of the thrashing engine to be forwarded to the President of the Board of Agriculture, which I shall attend to. He also says that he has sent my letter to the Navy Board, in hopes that the experiment of propelling vessels by steam may be tried under its sanction and expense.

"Perhaps it might be proper to wait the answer of the Navy Board before writing to Mr. Praed about propelling the canal boats. I am very much obliged to you for writing to Captain Gundry, about the Wheal Friendship engine. I expect to have a portable steam-whim and stamps at work at my own expense in a few days, which will prove for itself its utility; that being the only way to introduce new things. I would be very much obliged to you to say if Mr. Halse is to pay me for my past attendance at St. Ives about the breakwater. Enclosed you have a letter to Sir John Sinclair, unsealed for your inspection, which, if you approve of, please to forward.

"I remain, Sir,

"Your very humble servant,

"RICHARD TREVITHICK."

Trevithick's skill did not prevent his being reasonably modest, or cause him to be envious of others; neither did his dear-bought experience, that one's own pocket must pay for making public one's own inventions, prevent his again soliciting the assistance of persons of influence, though it does not seem that

Mr. Praed helped forward the screw-propeller, or that Sir John Sinclair gave direct help, though he probably made known the high-pressure steam-engine to the marine experimenters on the Clyde.

[Rough draft.]

" Mr. Rastrick,   " Camborne, *December 7th*, 1812.

"Sir,—I have been waiting your answer to my last, and especially that part respecting the West India engine, as I have not nor could not answer their letter to me without first hearing from you; therefore must beg you will be so good as to answer me by return of post on that subject. If they get impatient about the time, and refuse to take the engine, I have no doubt the Plymouth people will take it and several others; but I very much wish to send one to the West Indies, as there is a large field open there for engines of this kind. I have received an order for a thrashing engine for Lord de Dunstanville, of Tehidy; and as I wish those thrashing engines to be known through the country, I intend to take one of the engines ordered for Padstow and send it to Tehidy. One of the Padstow farmers can wait until you make another for him. Therefore I would thank you to send the first finished by ship from Bristol for Portreath or Hayle. Send a drum with everything complete, of which you are a better judge than I. Probably about 3 feet in diameter and 3½ feet long will be sufficient.

"There must be a fly-wheel with a notch to carry the rope, and also a small notch-wheel on the drum-axle. I think 6½ feet diameter for the fly, and 9½ inches diameter for the small wheel, will give speed enough to the drum. Mind to cast a lump, or screw on a balance, of about 1 cwt., on one side of the fly-wheel. There must be two stands on the boiler, and a crank-axle, or otherwise a crank-pin, in the fly-wheel, whichever you please; with a shaft 3 feet long with a carriage.

"The engine is to stand in a room under the barn, about 7½ feet high, 7 feet wide, and 14 feet long. The fly-wheel will stand across the narrow way of the room. The rope will go up through the floor, and the drum be shifted by a screw, horizontally, on the barn floor, so as to tighten the rope. I shall put down the top of the boiler level with the surface, with an arched way to the fire and ash-pit under ground, to prevent the chance of fire, which the farmers are very much afraid of. I send you a sketch showing how it is to stand.

" I do not bind you to the size of the drum or wheels, only the room that the fly-wheel works in is but 7 feet wide.

" Now to Mr. Richards' mill.

" Query 1st.—The length of the piston, and the small variation that the beam will give it, is so trifling that it will not be felt.

"The cylinders that have been working on their sides for seven years past, are now working as well as any engine with upright cylinders, which is a proof that the little rubbing is of no consequence.

" Query 2nd.—The passage in the cock is equal to the passage we make in our large engines, which is only one-fortieth part of the piston; and as we shall work with high steam, we do not mind the pressing through the steam-passage; and as the steam will be very much expanded, it will not be felt in the passage to the condenser. I know where we have removed cylinders and put larger ones on the same nozzles and condensing work, and the engines did good duty.

" Query 3rd.—I find by experience that if you give double the quantity of injection to an engine one stroke, and none the other, that the quicksilver in the gauge will stand nearly the same ; the cold sides of the condenser are sufficient to work an engine a great many strokes without any injection.

" Query 4th.—You may put a hanging to the air-pump bucket, and foot-valve ; either that or a rising one will do very well, but I think the rising cover and wood face on the top, best.

" Query 5th.—The air-pump bucket is large enough. At Wheal Alfred they have a 64-inch cylinder; the air-pump is 20 inches, and the stroke is half that of the engine. They were afraid that it was too small; they then put another of 14 inches by the side of the first, the same stroke. The quicksilver tube stands as high with the one 20-inch bucket as with the two buckets; the engine works best with the one bucket. I have found by experience that size to be sufficient, and (especially in an engine that works quickly) make the cistern high enough to cover the condensing work well with water.

" Query 6th.—My reason for making the forcing pump with duck-valves is, because they do not bum like the others, and we find them seldom out of repair ; but make it whichever way you think best, and work it in any way you like.

" Query 7th.—I mean by ¾ expansive, that the steam is to be shut off from the cylinder when the piston has moved up from the bottom one quarter of its stroke. Make the cam to your own mind.

" Query 8th.—I do not think the engine will require a heavier fly-wheel, as the stones will act as a fly, and the power, though so very irregular, will be so sudden in its changes, that the speed of the machinery will not let it be felt. If you make a crank, you may make the fly-wheel 3 or 4 feet more in diameter. But if with a pin in the fly-wheel, the beam would come down on the top of it; therefore, I think it will be better to put a crank, and put the fly-wheel in the middle of the shaft.

" Query 9th.—The steam will be raised to 25 lbs. to the inch above the atmosphere, or 40 lbs. to the inch on a vacuum; but I think you need not calculate for much more strength on that account. It is not the power that breaks the machinery, but bangs, and not the uniform weight that this will give.

" Query 10th.—Twenty strokes per minute I propose, which I think a fair speed.

" Query 11th.—The fire-bars must be of wrought iron; we find them answer much better than cast iron. Let them be ⅝ths of an inch from bar to bar, 1 inch thick at the top, ⅜ths of an inch at the bottom, 2 inches deep, 4 feet long, with bits on them at the ends, to prevent their getting too close together. I find the nearer the fire is to the door, the better and handier it is to work. All the large engines are in this way, and we do not find the door or front plate get hot, as they are lined with brick. Cast the door with a rib to hold a brick on its edge. Tube, 2 feet 9 inches by 1 foot 11 inches; manhole, 15 by 10 inches.

" Query 12th.—A governor will be required ; perhaps as good a place as any for it, out of the way, will be on the cast iron that carries the beam ; you may turn the fly-wheel whichever way you please. If this engine is worked with steam of 25 lbs. to the inch above the atmosphere, and the steam shut off at one-twentieth part of the ascending stroke of the piston, the power will be as three is to two of Boulton and Watt's single engines.

" Only two pairs of stones for the present, but calculate those stones to stand in such a way that another pair may be placed, on a future day, if wanted. I have not seen Mr. Richards lately. I wish you to write a form of an order, in your next, such as you wish, and I will get him to write to you accordingly. Put the engine and drum for Lord de Dunstanville out of hand neat and well, as it will be well paid for ; and make the stands, &c., in your own way.

" RD. TREVITHICK."

Mr. Richards' flour-mill engine may claim to be the first practical smoke-burner: keeping the fire much thinner at the inner end of the grate-bars than at the fire-door end of the grate, allowed of the freer passage of air through the thinner layer of coal, near the fire-bridge, causing the combustion of the passing gas. This idea has, since the date of Trevithick's letter, led to several smoke-burning patents. The boiler fire-tube was oval, 2 feet 9 inches by 1 foot 11 inches. The open-topped cylinder was supplied with a heavy and deep piston serving as a counterweight, and also as a guide in the cylinder for correcting the angle of the connecting rod. Experience had taught him that the cold sides of the condenser were sufficient to work an engine a great many strokes without a supply of injection ; and he had already used high-pressure steam of 25 lbs. to the inch above the atmosphere, cut off from the cylinder when the piston had performed one quarter

of its course : thus both these things were as first steps
leading to the modern expansive steam-engine and surface condensation.

The simplicity of the engine is remarkable—a high-
pressure, expansive, condensing engine, worked by a
single four-way cock, without cylinder-cover, or parallel
motion.

The low first cost, and non-liability to derangement,
were always kept in view; and his confirmed experience in the satisfactory working of horizontal cylinders
prior to 1812 illustrates their extended application ;
for at that time scarcely any other engineer had constructed other than upright cylinder engines. No detail
escaped his observant gaze. The fire-bars were to be
2 inches deep, 1 inch thick at the top edge, tapered to
$\frac{3}{4}$ths of an inch at the bottom, giving the required
strength, with free room for air, which in its passage
cooled the bar, carrying the heat into the fire. Years
before and after that period the fire-bar in common use
by thoughtless people was a square iron bar that was
always burning and bending.

The letter is descriptive of the high-pressure steam-
engine in the sixteenth year of its age; and its expansive steam, made practical by Trevithick's high-pressure
boilers. This engine only took steam during the first
quarter of its stroke, the remaining three-quarters were
by the expansion. Had it taken steam only during one-
twentieth of its stroke, it would have been more powerful than Boulton and Watt's low-pressure steam vacuum
engine of the same size.

[Rough draft.]

" SIR,   " CAMBORNE, *November 8th*, 1812.

    " I have your favour of the 3rd inst., informing me that
Messrs. Fox and Williams have engaged to quarry the stone for

the breakwater at Plymouth, but does not say whether you hold
any share with them in the contract or not. Therefore I cannot
understand from your letter whether you wish to see an engine
fitted to the purpose of the breakwater, or for pumping the
water from the foundations of the Exeter Bridge. Please to
inform me which of the two purposes you wish to see the engine
calculated for, and about what time you think you shall want it,
and I will get one finished suitable to the purpose you intend
it for.

" Yours, &c.,

" Jas. Green, Esq., *St. David's Hill, Exeter.*   " R. T.

" N.B.—To what extent have Messrs. Fox engaged, and what
parts of the work do they perform? I think more good might
be done by loading, carrying, and discharging, than by quarrying only."

Trevithick was equally ready with the application of
steam-power either for pumping of water or for boring
and removing rock. The use of chisels and rock-
breakers in the Thames in 1803[1] had prepared the
way for the more perfect engine for boring, lifting,
and carrying rock from the quarries to its destination
at the Plymouth Breakwater in 1812.[2]

" Sir,   " 106, Holborn Hill, *November 26th*, 1812.

" I am in receipt of yours of the 22nd inst. Mr. Giddy
informs me that Mr. Fox and Mr. Williams are to have 2s. 6d.
per ton for making the breakwater at Plymouth, and he con-
siders that they can do it for 2s., which he thought would give
them 50,000l. profit. If you meet those gentlemen, I have to
caution you not to LEARN THEM anything until you make a
bargain, as I know Mr. Williams will endeavour to learn all he
can and then you may go whistle.

" If 6d. per ton will give 50,000l. profit, a halfpenny per ton
would give upwards of 4000l. Would they agree to give you
that for your labour only? However, this will depend in a great

---

[1] See Stonebreaker of 1803, vol. i., p. 239.
[2] See Steam-crane, vol. i., pp. 162, 274.

measure on the time it will take in doing. If it takes eight years it would be 500*l.* a year for you (according to Mr. Giddy's calculation).

" Your well-wisher,

" HENRY HARVEY.

"MR. RD. TREVITHICK, *Camborne.*"

Mr. Harvey knew Trevithick's weakness in money matters. Rennie had been employed to report on the proposed Plymouth Breakwater, and in 1811 was desired by Lord Melville, the head of the Admiralty, to proceed with the work. "The price paid in 1812 for taking and depositing rubble in the breakwater was 2*s.* 9*d.* per ton; it was afterwards reduced to 1*s.* per ton. A piece of ground was purchased from the Duke of Bedford at Oreston, up the Catwater, containing 25 acres of limestone, well adapted for the purposes of the work; and steps were taken to open out the quarry, to lay down railways to the wharves, to erect cranes."[1] The idea of the plan to be followed in conveying stone with greater economy and dispatch than was contemplated by Rennie, originated with Trevithick, while the former received the credit and the pay, as he before had done with the steam-dredger.

[Rough draft.]

"MR. Fox, Jun.,   "CAMBORNE, *January* 29*th*, 1813.

"Sir,—Since I was at Roskrow I have been making trial on boring lumps of Plymouth limestone at Hayle Foundry, and find that I can bore holes five times as fast with a borer turned round than by a blow or jumping-down in the usual way, and the edge of the boring bit was scarcely worn or injured by grinding against the stone, as might have been expected. I think the engine that is preparing for this purpose will bore ten

---

[1] 'Lives of the Engineers,' by Smiles, vol. ii., p. 260.

holes of 2½ inches in diameter 4 feet deep per hour. Now suppose the engine to stand on the top of the cliff, or on any level surface, and a row of holes bored, 4 feet in from the edge of the cliff, 4 feet deep, and about 12 inches from hole to hole for the width of the piece to be brought down at one time, and wedges driven into the holes to split the rock in the same way as they cleave moorstone, only instead of holes 4 inches deep, which will cleave a moorstone rock 10 feet deep when the holes are 14 or 15 inches apart, the holes in limestone must go as deep as you intend to cleave out each stope, otherwise the rock will cleave in an oblique direction, because detached moorstone rocks have nothing to hold them at the bottom, and split down the whole depth of the rock. In carrying down a large piece of solid ground the bottom will always be fast, therefore unless it is wedged hard at the bottom of the hole the stope cannot be carried down square. In a hole 2½ inches diameter and 4 feet deep put in two pieces of iron, one on each side of the hole, having a rounded back, then put a wedge between the two pieces, which might be made thus, if required to wedge tighter at the bottom of the hole than at the top.

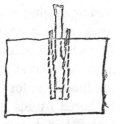

"If this plan answers, the whole of the stones would be fit for service, even for building, and would all be nearly of the same size and figure. Each piece would be easily removed from the spot by an engine on a carriage working a crane, which would place them into the ship's hold at once. It would all stand on a plain surface, and might be had in one, two, three, or four tons in a stone, as might best suit the purpose, which would make the work from beginning to end one uniform piece. Steam machinery would accomplish more than nine-tenths of all the work, besides saving the expense of all the powder. I find that limestone will split much easier than moorstone, and I think that a very great saving in expense and time may be made if the plan is adopted.

"Please to think of these hints and write me when and where I may see you to consult on the best method of making the tools for this purpose before I set the workmen to make them. Any day will suit me, except Monday, the 8th of February. The sooner the better, as I cannot set to work to make the tools until we have arranged the plan.

" I am, Sir,

" Your humble servant,

" Rd. T."

The successful completion of the Mont Cenis Tunnel in 1871 was mainly due to an ingenious application of combined mechanical force to boring tools, before limited to man's strength; but the applied principle existed sixty years ago, and though not so perfect in detail, yet more comprehensive. Trevithick's high-pressure steam boring engine enabled him to penetrate the rock five times as fast as the quarryman's power. Ten holes, $2\frac{1}{2}$ inches in diameter, 4 feet deep, could be bored in an hour, and he sagaciously suggested that in quarrying the limestone for the breakwater, a row of holes should be bored by his engine 4 feet in from the face of the rock, $2\frac{1}{2}$ inches in diameter, 4 feet deep, and 12 inches apart; and by dropping into each hole two half-round pieces of iron, to be driven asunder by a steel wedge, large blocks would be forced off without the use of gunpowder. The high-pressure steam-puffer having bored the stone, moved itself toward the broken mass, lifted it into waggons, and again changing its powers from steam-crane to steam-locomotive, conveyed it to the port, and lifted it into the ship's hold. The whole operation was thus aptly described by the inventor, who then counted on contracting for the breakwater work:—" Steam machinery will accomplish more than nine-tenths of all the work, besides saving the expense of all the gunpowder."

[Rough draft.]

" MR. ROBERT FOX, Jun.,   "CAMBORNE, *February 4th*, 1813.

"Sir,—Since I was with you at Falmouth I have made a trial of boring limestone, and find that the men will bore a hole 1½ inch in diameter 1 inch deep in every minute, with a weight of 500 lbs. on the bit. I had no lump more than 12 inches deep; but to that depth I found that having a flat stem to the bit of the same width as the diameter of the hole, twisted like a screw, completely discharged the powdered limestone from the bottom of the hole without the least inconvenience.

"From the time the two men were employed boring a hole 12 inches deep, I am convinced to a certainty that the engine at Hayle will bore as many holes in one day as will be sufficient to split above 100 tons of limestone, and would draw that 100 tons of stone from the spot and put them into the ship's hold in one other day. The engine would burn in two days 15 bushels of coal, four men would be sufficient to attend on the engine, cleave the stone, and put it into the ship's hold. I think it would not amount to above 9d. per ton, every expense included, but say 1s., which I am certain it will not amount to. Perhaps it may not be amiss to withhold the method of executing this work until the partners have more fully arranged with me the agreement as to what I was to receive for carrying the plan into execution. I do not wish that anyone but your father should be made acquainted with the plan, and have no doubt he will have sufficient confidence in the scheme to adopt it. I shall be glad to hear from you soon, as I intend to go to Padstow in a few days and shall not return under a fortnight.

"Your humble servant,

"RD. TREVITHICK.

"N.B.—I this day received a letter from Mr. Gould, requesting to know what the expense of an engine and apparatus would be for clearing Falmouth Harbour, which I have sent by the post." [1]

---

[1] See letter, 4th February, 1813, vol. i., p. 248.

It had been and still is the custom to bore rock either with a long and heavy jumper-chisel, lifted a foot or two, and falling by its own weight, pounding to powder a portion of the rock, or by the use of a much smaller chisel called a borer, struck by a hammer. Trevithick having made his steam-engine perform those jumper and borer movements, turned his attention to the improvement of the borer, and found that a revolving bit was more suitable for drilling limestone than the borer-chisel. The powdered stone was removed from the hole by giving a screw form to the stem of the bit. Many years afterwards precisely similar bits for boring wood were patented as new things, and are still used. Within five months of his first communication with the contractors for the Plymouth Breakwater he had designed and made an engine to bore, lift, and convey to the ship's hold 50 tons of stone daily at less than half the cost Rennie was then paying for it.

[Rough draft.]

" SIR,   " CAMBORNE, *February 4th*, 1813.

  " I have your letter of the 31st January requesting to know the time in which the engine will be ready for the bridge at Exeter, and also about giving an additional power to it.

  " The engine shall be ready in six weeks from the end of January, and shall be capable of lifting the 10-inch bucket you have ordered instead of the 9-inch before proposed, which was to have delivered 500 gallons of water 12 feet high per minute; but now the engine shall be made to lift in the same proportion as a 9-inch is to a 10-inch bucket, which will be 617 gallons of water per minute instead of 500 gallons, as was before agreed on, and I shall charge you accordingly. I observe that you have ordered the pump, and from the description you give of it, I think it will answer very well. If you wish a perpendicular cylinder instead of a horizontal, I can construct it in that way, but it will not be so convenient for a portable engine. I have now engines with horizontal cylinders at work above ten years, and find them answer equally as well as a perpendicular cylinder.

"I remain, Sir,

"Your very humble servant,

"RICHD. TREVITHICK.

"JAS. GREEN, Esq., *Exeter.*"

Engineers nowaday are not in the habit of designing and constructing a steam-engine in six weeks, or willing to alter the agreed form from the horizontal to the vertical without charge.

[Rough draft.]

"Mr. Robert Fox, jun., informed me the other day that you had the sole direction of the work at Plymouth. Had I known it at the time you were at Scorier I should have communicated to you my ideas relating to the application of machinery there; but until a few days since I had an idea that the young Mr. Fox was about to take an active part in the management, which I now find was never his intention, only he very much wishes to have an experiment tried to see to what extent an engine was capable of performing as against men. An engine is now preparing for that purpose."

"SIR, "CAMBORNE, *February 24th*, 1813.

"On my return from Padstow this evening, where I have been for the last fortnight, I found your letter of the 11th inst. respecting the getting an apparatus ready for the Plymouth undertaking. Before I set about it I wish to see you and Mr. Fox, and will call any day you may appoint. Waiting your reply,

"I remain, Sir,

"Your very humble servant,

"R. TREVITHICK.

"MR. ROBERT FOX, Jun., *Fulmouth.*"

After three months of experimental scheming, without a thought of keeping his inventions secret, Trevithick for a moment became worldly wise, and asked for a written agreement before sending his locomotive boring engine to the breakwater.

[Rough draft.]

" Mr. Fox,   " CAMBORNE, 14th March, 1814.

" Sir,—I expect to be called to London immediately after the end of this month. The engine with the boring apparatus for Plymouth remains at Redruth. I very much wish to see you on that business before I leave home, and would be much obliged by your dropping me a note by post, saying what day it would be convenient for me to wait on you.

" R. T."

The rock-boring machine was completed, and reached the breakwater two months after his interview with the Foxes, who were prominent in the quarrying work. " The engine for Plymouth will be put to break the ground as soon as I can find time to go up there."[1] It was impossible for any one man, single-handed, to make perfect such numerous practical inventions as were undertaken by Trevithick at that time. His letter of a few months before[2] reveals the facility with which he moulded the steam-engine to his requirements. " The ploughing engine that I sent you a drawing for, after being used for that purpose, was to have been sent to Exeter for pumping water. I have been obliged to take the small portable engine from Wheal Alfred Mine, and have a new apparatus fitted to it for Plymouth Breakwater. A small engine which I had at work at a mine I have been obliged to send to the farmers for thrashing." Messrs. Fox would probably require many engines for the Plymouth Breakwater,

---

[1] Trevithick's letter, May 20, 1813, chap. xxi.
[2] See letter to Mr. Rastrick, January 26, 1813, chap. xviii.

having engaged with Government to deliver three million tons of stone; and to prevent delay, the boring apparatus was applied to an engine made for another purpose, while drawings for a new and more suitable engine for boring stone were sent to Mr. Rastrick.

He engaged that an engine should bore holes to split 100 tons of limestone a day; and that on the following day it should, as a locomotive and steam-crane, load that quantity in waggons, convey it from the quarry to the port of shipment, and then by steam-crane place it in the hold of a vessel. The whole of the work to be done by 11 cwt. of coal and four men. The gross cost would be 1s. per ton for breaking and removing, though at that time Mr. Rennie was paying 2s. 9d. a ton, which in after years was reduced to 1s., just what Trevithick said was a fair price.

While this ready application of the high-pressure steam-engine was going on in England, it had also extended to, and was coining money in the Mint at, Lima, where Trevithick contemplated going to look after it, intending to land at Buenos Ayres, and make his way across the continent of South America and the mountains of the Cordilleras as best he could, leaving the home field he had made so fertile to be reaped by others, and the stone-boring locomotive to be forgotten for many years.

[Rough draft.]

" SIRS,   " PENZANCE, *December 9th*, 1815.

" Your very great neglect in not writing . . . . Herland engine will work, I expect, in about fifteen days. It is a plunger of 33 inches diameter, 10-feet stroke, with a double packing around the top of the plunger-pole, in the same way as the steam is turned into the stuffing box of a double engine to exclude the air, only there is a small tube from the bottom of the boiler to the middle of the stuffing box to prevent the escape of steam.

"I am sorry to find by Mr. Uville's letter that the Mint engine does not go well. I wish you had put the fire under the boiler and through the tube, as I desired you to do, in the usual way of the long boilers; then you might have made your fire-place as large as you pleased, which would have answered the purpose and worked with wood just as well as with coal. I always told you that the fire-place in the boiler was large enough for coal, but not for wood; also if you found that the cock did not open and shut in proper time, to make the gear to it work the same as the Dolcoath puffer whim-engine instead of the circular gear. The boiler is strong enough and large enough to work this engine with 30 lbs. to the inch, thirty strokes per minute. I hope to leave Cornwall for Lima about the end of this month, and go by way of Buenos Ayres, and cross over the continent of South America, because I cannot get any other passage. None of the South Sea whalers will engage to take me to Lima, as they say they may touch at Lima or they may not. Unless I give them an immense sum they will not engage to drop me there. To be brought back to England after a two years' voyage without seeing Lima would be a very foolish trip. To make a certainty I shall take the first ship for Buenos Ayres, preparations for which I have already made."

This unfinished rough draft was intended for one of the men who had gone to Lima, less fruitful in emergency than Trevithick, who, without a moment's hesitation, would have constructed a fire-place outside the boiler, when the internal tube fire-place was found to be too small for a wood fire. Trevithick's proposing sixty years ago to make his way over the almost unknown track from Buenos Ayres, on the Atlantic, to Lima, on the Pacific, was perhaps characteristic of his daring spirit, that turned all things to good account; but he dreamed not that his grandson and namesake would at this time be conducting the steam-horse on the

same line of march on the Central Argentine Railway
from Rosario to Cordoba, in the Argentine Republic.

[Rough draft.]

" GENTLEMEN,   " CAMBORNE, *August* 19*th*, 1813.

" I received yours at Bridgenorth of the 19th July,
ordering a steam-engine for rolling sugar-cane. I immediately
set the founders to work on one for you, which is to be ready
by my return to Bridgenorth about the end of September. I
intend to ship it for Bristol, and will call on you on my journey
down to Cornwall, as I intend to set it to work at Bristol for
your inspection before it is put on board ship. The price I
cannot accurately say at present, as the engine now making is
on a new principle ; and as it will be more simple in construction,
I hope to be able to render it within the price before stated to
you. As it is on a new plan I cannot fix the price until I know
the cost of making. All I can say at present is that it shall not
exceed what I stated to you in my former letter.

" I remain, Sirs,

" Your humble servant,

" RD. TREVITHICK.

"MESSRS. PINNEYS AND AMES, *Bristol.*"

The engine for the sugar plantations in Jamaica, on
an improved plan, was to be constructed in the short
space of six weeks, and if a saving in cost was effected,
the inventor would hand the whole of it to the purchaser.

[Rough draft.]

" SIR,   " PENZANCE, CORNWALL, *March* 8*th*, 1816.

" I received your favour of the 25th January, but did not
answer it in due course, because I was then erecting a very large
engine, which is the first on a new plan. This engine, which
has been at work about a month, performs exceedingly well.
The cost of erection and the consumption of coal are not above
one-third of a Boulton and Watt's, to perform the same work.

An engine of 4-horse power will not require a space of more than 5 feet high, 5 feet long, and 3 feet 6 inches wide. In some instances I employ a balanced wheel 5 feet in diameter. The water required will be a pint and a half per minute. The coal, one quarter of a bushel or 21 lbs. per hour. The price of a machine, finished and set to work, 100 guineas. It does not require either wood or mason work, but stands independent of every fixture, and may be set to work in half an hour after being brought on your premises.

"Your obedient servant,

"RICHD. TREVITHICK.

"DR. MOORE, M.D., *Exeter.*"

A 4-horse-power portable high-pressure steam-puffer engine cost 105*l.*, with internal fire-tube and machinery attached to the boiler, ready for work in half an hour after lighting the fire, consumed 21 lbs. of coal and 11 gallons of water for each hour's work, at a cost of threepence.

The reader's attention has been very imperfectly drawn to the numerous subjects touched on in these remnants of Trevithick's correspondence between the years 1804 and 1816; among them may be traced the portable high-pressure steam-engine, the tubular cylindrical boiler of wrought iron, the economy of expansive working with steam of 100 lbs. on the inch, but limiting it to 20 lbs. when not in the charge of experienced workmen, and testing boilers by water pressure to four times the intended working pressure.

The economy of heat in smelting furnaces and in the aërated steam-engine were bold means to large results. The cheap 100*l.* steam-engine of 1812, with open-top cylinder and rigid simplicity of gear, resembling Newcomen's first atmospheric engine, was really a high-pressure steam expansive engine, with the germ of

surface condensation, as ready to convey itself from
mine to mine or from farm to farm, and to join in
performing labourer's work, even to boring and conveying rock by land or sea, as the most perfect of
modern engines; and yet this unadorned engine, as
seen in the agricultural engine of the following chapter,
followed the excellent mechanism of the double-acting
Kensington model of 1798, and the still more beautiful
engine of the 1802 patent and London locomotive.

CHAPTER XVIII.

AGRICULTURAL ENGINES.

THE late Mrs. Trevithick said "that during the difficulties in London in 1808 and 1810, when Trevithick was overwhelming himself with new experiments and the cost of patents, and law expenses, lawyers and bailiffs took everything worth having from her house, including account-books, drawings, papers, and models, which she never saw again."

His earlier account-books left in safety in his Cornish home, though very disconnected, give trustworthy traces of his work up to 1803. From that time only detached accounts or papers are found until 1812, when the unused pages in two old mine account-books of his father served as his letter (rough-draft) books; and judging from their number and style, his correspondence was most extensive and varied.

[Rough draft.]

" HAYLE FOUNDRY, *February* 13*th*, 1812.

" To SIR CHRISTOPHER HAWKINS, Baronet.

"Sir,—I now send you, agreeable to your request, a plan and description of my patent steam-engine, which I lately erected on your farm for working a thrashing mill. The steam-engine is equal in power to four horses, having a cylinder of 9 inches in diameter. The cylinder, with a moderate heat in the boiler, makes thirty strokes in a minute, and as many revolutions of the fly-wheel, to every one of which the drum of the thrashing mill (which is 3 feet in diameter) is turned twelve times. The boiler evaporates 9 gallons of water in an hour, and works six hours without being replenished. The engine requires very little attention—a common labouring man easily regulates it.

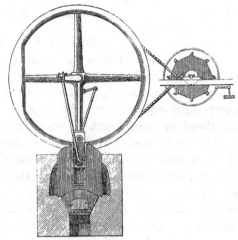

TREVITHICK'S HIGH-PRESSURE STEAM-PUFFER THRASHING ENGINE, 1812.

" The expense of your engine of 4-horse power, compared with the expense of four horses, is as follows:—

	£	s.
Original cost of the steam-engine	80	0
Building material and rope	10	0
	£90	0
Interest on the above 90l. at 5 per cent.	4	10
Wear and tear at 5 per cent.	4	10
	9	0
Original cost of horse machinery for four horses	£60	0
Interest on the above at 5 per cent.	3	0
Wear and tear at 15 per cent.	9	0
	12	0

" Two bushels or 164 lbs. of coal will do the work of four horses, costing 2s. 6d.

" Four horses at 5s. each, gives 20s.   Cost of coal, 2s. 6d. as compared with 20s. for horses.

" I remain, Sir,

" Your obedient servant,

" RICHARD TREVITHICK."

" CORNWALL, *February 20th,* 1812.

" Having been requested to witness and report on the effect of steam applied to work a mill for thrashing corn at Trewithen, we hereby certify that a fire was lighted under the boiler of the engine five minutes after eight o'clock, and at twenty-five minutes after nine the thrashing mill began to work, in which time 1 bushel of coal was consumed.   That from the time the mill began to work to two minutes after two o'clock, being four hours and three-quarters, 1500 sheaves of barley were thrashed clean, and 1 bushel of coal more was consumed.   We think there was sufficient steam remaining in the boiler to have thrashed from 50 to 100 sheaves more barley, and the water in the boiler was by no means exhausted.   We had the satisfaction to observe that a common labourer regulated the thrashing mill, and in a moment of time made it go faster, slower, or entirely to cease working.   We approve of the steadiness and the velocity with which the machine worked; and in every respect we prefer the power of steam, as here applied, to that of horses.

(Signed)   " MATTHEW ROBERTS, Lamellyn.

" THOMAS NANKIVILL, Golden.

" MATTHEW DOBLE, Barthlever."

This first high-pressure steam thrashing machine was working on the 13th February, 1812, at Trewithen, the property of Sir Christopher Hawkins, as proved by Trevithick's drawing of the machine, his account of the work performed, and the report of the three wise men that the power of steam was preferable to the power of horses.   Its first cost was less than that of a horse machine; but to make the calculated amounts come right Trevithick charged 15 per cent. for wear and tear on the horse machinery, and but 5 per cent. on the steam-engine; overlooking the cost of the horses, which would have made the outlay for the horses and machinery greater than for the steam-engine.

The whole design evidences simplicity and consequent cheapness; no complication of valves or valve-gear, no cylinder cover, parallel motion, guide-rods, or air-pump, with its condenser and injection-water.

The 4-horse engine, with boiler complete, cost 90*l.* A common labourer worked it, and as it needed no supply of feed-water during six hours of work, the cost and attention of supplying feed were avoided. If a supply was required during the day it could be given by a pipe with two taps.

This first use of steam in agriculture was immediately followed by Lord Dedunstanville of Tehidy, Mr. Kendal of Padstow, and Mr. Jasper of Bridgenorth. Sir Charles' request for a more official report signed by disinterested persons brought a reply that the thrashing engine continued to work well. "It far exceeds my expectation. I am now building a portable steam-whim, on the same plan, to go itself from shaft to shaft." "If you should fall in with any West India planter that stands in want of an engine, he may see this at work in a month, which will prove to him the advantage of a portable engine to travel from one plantation to another. The price complete is 105*l.*"[1]

"DEAR SIR,   "ARGYLE STREET, 19*th March*, 1812.

"I am sorry it is not convenient for me to advance you the money for Wheal Liberty; adventurers having the dues very low, ought to furnish the needful. I am very glad you have

---

[1] See Trevithick's letter, 10th March, 1812, chap. xx.

succeeded with your portable steam-engine, and am persuaded they will be more and more adopted. I have shown your account of your thrashing by steam, and Sir John Sinclair and Mr. —— very highly approve it. Sir John Sinclair wished the communication had been made to the Board of Agriculture. Sir John wished me to transmit you the enclosed on coals moved by steam. .... whether you had a plan of this sort, as they would be very serviceable in passing the friths in Scotland. He seems to think you ought to advertise your steam-engines for thrashing; indeed, I think so too.

"By the enclosed letter, Sir John Sinclair wishes you to send him an account of your improved steam-engines. You will be careful in drawing up your letter to Sir John, because it will probably be read to the Board of Agriculture, and perhaps inserted in their publication. You will begin by acknowledging his letter, of date .... respecting the American passage boat .... and your improved small steam-engine. You will give him an account of the saving you have effected at Dolcoath, and a certificate of the same by the mining captains; the engine for thrashing you built for me, and the work it did, and the coals it consumed; the expense of the steam-engines, and the uses they may be applied to.

"I remain, dear Sir,
"Yours most obediently,
"C. HAWKINS."

In 1812 Trevithick advertised the use and sale of steam-engines, weighing 15 cwt., costing 63*l.*, for thrashing, grinding, sawing, or other home work; and also a more powerful engine for the steam-plough, or the harrow and spade machine for 105*l.*, to travel from farm to farm. He wrote to Sir John Sinclair:[1]—

"I received from Sir Charles Hawkins a copy of Dr. Logan's letter to you, also a note from you to Sir Charles Hawkins, both respecting the driving boats by steam; respecting the

---

[1] See Trevithick's letter, 26th March, 1812, chap. xv.

engine for thrashing, chaff-cutting, sawing, &c. I am now making one of about two-thirds the size of Sir Charles Hawkins', which will be portable on wheels. By placing the engine in the farm-yard, and passing the rope from the fly-wheel through the barn-door, or window, and around the drum on the machine axle, it may be driven.

" The steam may be raised, and the engine moved a distance of two miles, and the thrashing machine at work, within one hour.

" The weight, including engine, carriage, and wheels, will not exceed 15 cwt.; about the weight of an empty one-horse cart.

" The size is 3 feet diameter, and 6 feet high. If you wish to have one of this size sent to the Board of Agriculture as a specimen, the price delivered in London will be sixty guineas."

This engine differed from that referred to in the drawing of Sir Charles Hawkins, mainly in the boiler having the fire-place in the fire-tube, requiring no brickwork, and having the advantage of portability. It was very like the earlier locomotive boiler, except that it was placed upright, as steam-cranes now use boilers, instead of being horizontal.

[Rough draft.]

" CAMBORNE, CORNWALL, *April 26th*, 1812.
" To SIR JOHN SINCLAIR.

" I have your favour of the 4th instant, informing me that you had sent my letter respecting propelling ships by steam to the Navy Board; and also requesting a drawing and statement of the thrashing engine to be sent to the President of the Board of Agriculture, which shall be forwarded immediately.

" I beg to trouble you with a few wild ideas of mine, which perhaps may some future day benefit the public, but at this time remain buried, for want of encouragement to carry it into execution.

" The average consumption of coals in large steam-engines is about 84 lbs. (or one bushel), to lift 10,000 tons of water or earth 1 foot high.

" The average cost of this coal in the kingdom is sixpence. The average of a horse's labour for one day is about 4000 tons lifted 1 foot high, costing about 5s.

" A man's labour for one day is about 500 tons lifted 1 foot high, costing 3s. 6d.

" I have had repeated trials of the water lifted by coals, horses, and men, proving that where a bushel of coal can be purchased for sixpence, that sixpence is equal to 20s. of horse labour, and to 3l. 10s. of men's labour.

" If you calculate a man to lift 500 tons 1 foot high, it is equal to 100 tons lifted 5 feet high ; a very hard task for a man to perform in a day's work.

" This calculation proves the great advantage of elemental power over animal power, which latter I believe can in a great part be dispensed with if properly attended to, especially as we have an inexhaustible quantity of coals.

" To prove to you that my ideas are not *mere* ideas, in general my wild ideas lead to theory, and theory leads to practice, and then follows the result, which sometimes proves of essential service to the public.

" About six years ago I turned my thoughts to this subject, and made a travelling steam-engine at my own expense to try the experiment.

" I chained four waggons to the engine, each loaded with 2½ tons of iron, besides seventy men riding on the waggons, making altogether about 25 tons, and drew it on the road from Merthyr to the Quaker's-yard (in South Wales), a distance of 9¾ miles, at the rate of four miles per hour, without the assistance of either man or beast, and then, without the load, drove the engine on the road sixteen miles per hour.

" I thought this experiment would show to the public quite enough to recommend its general use ; but though promising to be of so much consequence, has so far remained buried, which discourages me from again trying, at my own expense, for the public, especially when my family call for the whole of my receipts from my mining concerns for their maintenance.

" It is my opinion that every part of agriculture might be performed by steam; carrying manure for the land, ploughing, harrowing, sowing, reaping, thrashing, and grinding; and all by the same machine, however large the estate.

" Even extensive commons might be tilled and effectually managed by a very few labourers, without the use of cattle.

" Two men would be sufficient to manage an engine, capable of performing the work of 100 horses every twenty-four hours ; requiring no extensive buildings or preparations for labourers or cattle, and having such immense power in one machine as could perform every part in its proper season, without trusting to labourers.

" I think a machine that would be equal to the power of 100 horses would cost about 500*l.*

" My labour in invention I would readily give to the public, if by a subscription such a machine could be accomplished and be made useful.

" It would double the population of this kingdom, for a great part of man's food now goes to horses, which would then be dispensed with, and so prevent importation of corn, and at a trifling expense make our markets the cheapest in the whole world ; because there are scarcely any coals to be found except in England, where the extreme price, duty included, does not exceed 2*s.* per bushel.

" I beg your pardon for having troubled you with such a wild idea, and so distant from being carried into execution ; but having already made the experiment before stated, which was carried out in the presence of above 10,000 spectators, who will vouch for the facts, I venture to write to you on the subject, for the first and only self-moving machine that ever was made to travel on a road, with 25 tons, at four miles per hour, and completely manageable by only one man, I think ought not to be dropped without further experiments, as the main point is already obtained, which is the power and its management.

" Your most obedient servant,

" RICHARD TREVITHICK."

The Board of Agriculture in 1812 had their attention drawn to the feasibility of using the steam-engine to save agricultural labour and lessen the cost of working land. Trevithick's intuitive knowledge told him his application would be in vain, though an engine was at work proving the saving of horse-power in the item of thrashing corn.

" I beg to trouble you with a few wild ideas of mine, which *perhaps may some future day benefit the public.*"

A steam-engine could exert as much power by the consumption of 6*d.* worth of coal as could be given by 20*s.* of horse-power, or by 70*s.* worth of men's power.

"Ideas lead to theory, theory leads to practice, then follows the result, which sometimes proves of essential service to the public."

"It is my opinion that every part of agriculture might be performed by steam. Carrying manure for the land, ploughing, harrowing, sowing, reaping, thrashing, and grinding; and all by the same machine, however large the estate."

"Two men would manage an engine capable of performing the work of 100 horses."

Such a use of the steam-engine, judiciously managed, " would double the population of this kingdom, and make our markets the cheapest in the world; because England is the country best supplied with coal and iron for steam-engines, and the land now growing food for horses would be available for man."

Its cost would be 500*l.*, and its power sufficient to propel the largest subsoil ploughs and tormentors; and had the Board of Agriculture supplied such a sum of money as is now ordinarily given by a farmer for a steam-plough, we should have had in 1812 ploughing, harrowing, sowing, reaping, &c., by steam. Years before, the same kind of engine had been made to work pumps, wind coal from shafts, drive rolling mills, tilt hammers, and steamboats, and convey material from place to place; and why should not his promise to the farmer be also made good with his increased knowledge derived from eight years of active experience? Receiving small encouragement in England, he applied to sugar-cane planters to give his engines a trial in the West Indies.

[Rough draft.]

"SIR CHARLES HAWKINS, Bart., "CAMBORNE, 1st *May*, 1812.

"Sir,—I have your favour of the 27th April, respecting a steam-engine for your friend for the West Indies, of the power of ten mules employed at one time. This power we calculate equal to forty mules every twenty-four hours, as six hours' hard labour is sufficient for one mule for one day.

"The expense of an engine of this power complete delivered in London would be 200*l.* The consumption of coals about 84 lbs., or one bushel, to equal the labour of three mules, or from 13 to 14 bushels of coal every twenty-four hours to perform the full work of forty mules (or in proportion for a lesser number), with a waste of about 15 gallons of water per hour, unless a reservoir was made to receive the steam, and then to work the same water over again.

"Where water is scarce, nearly the whole may be saved.

"You remarked that the rope might slip round the notch in the wheel; but to prevent any risk of that kind, I apply a small chain instead of the rope, which works the same as a chain on the barrel of a common thirty-hour clock.

"The speed of the periphery of the fly-wheel is about eight miles per hour, which I think is nearly double the speed of the mules when at work in the mill. This would reduce the size of the part which carries the chain on the cattle mill to half the diameter of the present walk of the cattle, which might be done without altering or interfering with the present cattle mills, and might, if required, either work separately or in conjunction with the mules in the same mill at the same time.

"To inform your friend of the power and effect of such an engine, I prefer his sending some person down to Cornwall, to see it tried on some of the cattle mills or whims in the mines.

"Engines that have been sent to the West Indies hitherto have cost nearer 2000*l.*; very large, heavy, and complicated machines, requiring 2500 gallons of water per hour for condensing, and could only be managed by a professed engineer, while any common labourer can keep in order and work these engines. If you prefer to send a person with it, the cost will be about 40*s*. per week.

"I remain, Sir,

"Your most obedient servant,

"RICHARD TREVITHICK."

This letter indirectly points out two long-standing radical errors in engineering phraseology. An early method of describing the value of an engine was by stating the number of pounds it would lift one foot high by the burning a bushel of coal, called the duty of an engine. Trevithick's bushel was 84 lbs., while other engineers, under the same term of bushel, meant various weights, up to 120 lbs.

Another form of speaking was the horse-power of an engine ; meaning that a horse could lift a certain number of pounds one foot high in a minute, and that a steam-engine lifting ten times as much was a 10-horse engine; but, as Trevithick points out, a horse only works at that rate for six hours out of twenty-four, while the steam-engine works continuously, performing the work of forty horses, yet is called a 10-horse engine.

The high-pressure engine suitable for the West Indies was to be adapted to the existing horse or mule machinery, that either power might be used. Its first cost and expense in working to be much less than that of the Watt low-pressure steam vacuum engine.

[Rough draft.]

"Sir Charles Hawkins, Bart.,   "Camborne, *June 13th*, 1812.

"Sir,—Yours of the 15th of last month I received, enclosing a drawing of a sugar-mill from Mr. Trecothick, which I should have answered per return, but was at that time in treaty for an engine for a sugar-mill with a Mr. Pickwood, who is in St. Kitts in the West Indies.

"The engine is now being erected at Hayle Foundry, of the power of twelve mules at a time, or equal to forty-eight mules during twenty-four hours.

"The cost is 210*l.* complete, with numerous duplicate parts.

"I hope she will be finished and sent off in a short time.

"I have now so fully proved the use of those engines, that I have engaged to take this one back if it does not answer their purpose, and to refund the whole sum if they return the engine to me in working order within four years.

"This gentleman says, if this engine answers he shall have two more for his own use, and four of his friends are waiting to see the result before ordering their engines.

"The mules that will be turned out of use by Mr. Pickwood's engine will sell for five times the sum the engine will cost him, exclusive of the wear of mules, with their keep and drivers, besides the greater dispatch and pleasantness of working a machine instead of forcing animals in so hot a climate.

"If your friend wishes an engine of this power and on the same terms, I can get two made and sent to London nearly in the same time as one. Enclosed I send to you a rough sketch of the engine and mill. I am of your opinion, that Sir John Sinclair has taken a useless journey by calling on the Navy Board, for nothing experimental will ever be tried or carried into effect except by individuals.

"If I could get an Act of Parliament for twenty-one years for only one-tenth part of the saving which I could gain over animal power and expense, I have no doubt but that I could get money to carry the plan fully into effect for propelling ships, for travelling with weights on roads, and doing almost every kind of agricultural labour.

"But a patent is but for fourteen years, and open to constant infringement; for the inventor of general and useful machinery is a target for every mechanic to shoot at, and unless protected or encouraged by some better plan than a common patent, will have the whole kingdom to contend with in law, and most likely receive ruin for his reward, which has too often been the case.

"A plan of such magnitude as this promises to be of, I think ought to be carried into effect by subscription, and as soon as accomplished, the subscribers to be repaid, and the invention thrown open for the use of the kingdom at large. I think about 1000*l.* or 1500*l.* would test the designs.

"It is expected that Mr. Praed will spend some time in this neighbourhood; I hope I shall be able to prove to you and to him the great use of propelling barges by steam. I have a small engine now at foundry, and would put it on board one of their barges for your inspection. I am very much obliged for your continued favours, and beg pardon for so often troubling you. I have so fully proved the great advantages resulting from those portable engines, that I very much wish the public to have the full use of them.

"I remain, Sir,

"Your most humble servant,

"RICHARD TREVITHICK."

A 12-mule-power engine for St. Kitts was being erected at Harvey's foundry at Hayle; Trevithick making himself liable for the whole cost, in case it should not answer the purpose. The mules thrown out of work by the engine would sell for five times as much as the engine cost, to say nothing of the saving in wear and tear of drivers and mules, and the unpleasantness of driving a mule in hot weather as compared with a machine.

If an Act of Parliament would give him one-tenth of the saving he could effect during twenty-one years, a company might be formed for carrying into full effect his plans for propelling ships, travelling with weights on roads, and performing almost every kind of agricultural labour, while a patent for fourteen years was open to constant infringement, and the inventor of useful machinery was a target for every mechanic to shoot at, had law suits with the kingdom at large, and ultimate ruin, as a reward for his labours. Inventions of such general application, when fairly established, should be thrown open to the public, Government paying the inventors their expenses, and reasonable reward for their time.

[Rough draft.]

"MR. PICKWOOD, "CORNWALL, CAMBORNE, 17th June, 1812.

"Sir,—Yours of the 17th April I received about twenty days since, and from that time to the present have been in treaty with Messrs. Plummer, Barham, and Co., for your engine. We have now closed for an engine complete, of the power of twelve mules at a time, with suitable duplicates, chains, &c., for 210l. I very much wish for your engine to be set to work by your own workmen, to show the planters the simplicity and easy management of the machine, and also save the expense of an engineer, which will tend to promote their use. The engine will be set to work before it is sent off, and every possible care taken to execute it in the most perfect order. From the experience I have had with common labourers keeping these engines in order, since I wrote to you, I have no doubt you will get on satisfactorily.

"I hope to get the engine ready in five or six weeks, but I fear there will be loss of time in shipping it. You may rest assured that I will spare no time or attention to promote the performance of this engine. I am so far satisfied with the probability of its fully answering your purpose that I voluntarily offered Messrs. Plummer, Barham, and Co., that if you return it to me for working repairs within four years, I will refund the whole of the sum I am to receive for it. I will take particular care to mark every part and send you a full description.

"Enclosed I send you a sketch of the engine attached to a sugar-mill. Please write to me by return of the packet; it may be in time, before the engine is shipped, to alter, or send you such things as I may not be acquainted with. I shall be glad to know the number of yards your mules travel in an hour when going at what you call a fair speed, in the mill, and also what number of rounds you wish the centre roll to make in an hour when worked by the engine.

"I remain, Sir,

"Your very humble servant,

"RD. TREVITHICK.

"R. W. PICKWOOD, Esq.,
"*St. Kitts, West Indies.*"

These are not the remarks of an uncertain schemer; every sentence having the impress of the ability and fixed intention of perfecting the work, and the belief that the simplicity of the engine would enable a common labourer to use it.

[Rough draft.]

"SIR CH. HAWKINS, Bart., "CAMBORNE, 5th July, 1812.

"Sir,—If your friend Mr. Trecothick intends to have a sugar-mill engine immediately finished and sent out with the one I am now making for Mr. Pickwood, he ought not to lose any time in giving his orders. I have made inquiry at Falmouth about sending out Mr. Pickwood's engine for St. Kitts on board a packet, which would save much time, but I fear it cannot be granted unless application is made by some person of note to the Post Office in London. Mr. Banfield of Falmouth told me that if application was made to send out a model as a trial, he had no doubt but it would be granted.

"This experiment with the portable engine that will travel from one plantation to another and work without condensing water, is certainly of the greatest consequence to the planters,

and as the whole weight will not exceed 1¾ ton, I should hope
that the Commissioners at the Post Office will grant this request.
I am sorry to trouble you so often about my business, but I beg
the favour of your goodness to inform me through what channel
I ought to make this application.

"I remain, Sir,

"Your most obedient humble servant,

"Rd. Trevithick."

This experiment with the 1¾ ton portable engine
to travel from one plantation to another, needing
no condensing water, was certainly of the greatest
consequence to the planters in the West Indies, and
should have been of equal importance to the people in
England.

Judging from the weight and cost, as compared with
agricultural engines of the present day, Trevithick was
nearer the mark then than we are now; its working
without condensing water the engineers of that day
believed to be impracticable, a fundamental error which
greatly retarded the use of the high-pressure steam-
engine.   The providing sufficient condensing water
was often a most serious item of cost, and as water
mains were not in use, a deep well was a necessary
part of a steam vacuum engine.

[Rough draft.]

"Gentlemen,   "Cornwall, Camborne, October 16th, 1812.

"Yours of the 30th of September I found at my house
on my return yesterday from a journey.   I am sorry to inform
you that Mr. Pickwood's engine is not ready.   Near three
months ago I set my smiths and boiler-makers to work to complete an engine for Mr. Pickwood, which parts were finished
five or six weeks ago.   The other parts of the engine, which
were to have been made of cast iron, were ordered and commenced at a foundry in this county, belonging to Blewett,

E 2

Harvey, and Vivian, and would have been finished and the
engine shipped long since had not these partners in the iron
foundry quarrelled with each other, and the Lord Chancellor
has laid an injunction and set idle their foundry. I have since
ordered the castings to be made at a foundry at Bridgenorth, in
Shropshire, belonging to Hazeldine, Rastrick, and Co., who will
complete the engine and send it to you in about two months, at
which time I intend to be in town to set it to work before it is
shipped for the West Indies.

" I remain, Gentlemen,

" Your very humble servant,

" RICHD. TREVITHICK.

"MESSRS. PLUMMER, BARHAM, AND CO.,
    " *London.*

" P.S.—Immediately on my receiving your order to prepare
an engine for Mr. Pickwood I wrote to inform him that I had
begun it, and enclosed a drawing of the engine with the method
of connecting the engine to the cattle-mills, and requested he
would remit to me his remarks on it, which I received by the
last packet, from which it appears for the best that the engine
is not in a forward state, because the parts would not have been
so suitable to the purpose as they will now be."

Fortune was against Trevithick. A difficulty between
his brother-in-law Harvey, and his old partner Vivian,
with Blewett, retarded the completion of the engine;
and the castings so anxiously waited for were ordered
from Hazeldine and Rastrick. The wrought-iron work
was made by the old smiths in his neighbourhood, who
had long been in the habit of hammering his schemes
into shape. This patchwork way of constructing
engines made success much more difficult.

Trevithick often laughed heartily at the following
incident which occurred during this quarrel at Harvey's works:—Blewett sent a handsome silver teapot
to Miss Betsy Harvey, who kept her brother's house,

called Foundry House. Trevithick was sitting with
them when the box was brought in and opened. Mr.
Henry Harvey was indignant at Mr. Blewett sending a
bribe or make-peace to his sister, and threw the silver
teapot under the fire-place. Trevithick, however,
quietly picked it up, pointed out the dinge it had received, wrapped his pocket handkerchief around it,
and saying, if it causes bad feeling here it will do for
Jane, marched away home with the pot. The writer
drank tea from it recently, and also laughed at the
dinge.

The following was written to Mr. Rastrick in December, 1812 : [1]—

"I have been waiting your answer to my last, and especially
that part respecting the West India engine, as there is a large
field there for engines of this kind. I have received an
order for a thrashing engine for Lord de Dunstanville of
Tehidy, and as I wish those thrashing engines to be known
through the country, I intend to take one of the engines
ordered for Padstow and send it to Tehidy; one of the Padstow
farmers can wait until you make another for him; therefore I
would thank you to send the first finished by ship from Bristol
for Portreath or Hayle. Send a drum with everything complete, of which you are a better judge than I; probably about
3 feet in diameter and 3½ feet long will be sufficient. There
must be a fly-wheel, with a notch to carry the rope, and also a
small notch wheel on the drum-axle. I think 6½ feet diameter
for the fly and 9½ inches diameter for the small wheel, will
give speed enough to the drum. Mind to cast a lump or screw
on a balance of about 1 cwt. on one side of the wheel. There
must be two stands on the boiler, and a crank-axle or otherwise
a crank-pin in the fly-wheel, whichever you please; with a shaft
3 feet long with a carriage. The engine to stand in a room
under the turn-about, 7½ feet high, 7 feet wide, and 17 feet

---

[1] See rough draft, Trevithick's letter, 7th December, 1812, chap. xvii.

long. The fly-wheel will stand across the narrow way of the
room. The rope will go up through the floor and the drum be
shifted by a screw, horizontally on the barn floor, so as to
tighten the rope. I shall put down the top of the boiler level
with the surface, with an arched way to the fire and ash-pit
underground to prevent the chance of fire, which the farmers
are very much afraid of.

"I send you a sketch showing how it is to stand. I do not
bind you to the size of the drum or wheels, only the room that
the fly-wheel works in is but 7 feet wide. Put the engine and
drum for Lord de Dunstanville out of hand neat and well, as it
will be well paid for, and make the stands, &c., in your own
way."

This description of Lord Dedunstanville's thrashing
machine illustrates the drawing of that supplied to Sir
Charles Hawkins.

[Rough draft.]

"MR. RASTRICK,   "CAMBORNE, *January 26th*, 1813.
"Sir,—I have your favour of the 10th inst., in which you
do not state the time when you expect I shall have either of
the engines that you are executing. As so much time has
elapsed since the orders were given, the persons that ordered
them are quite impotent. The ploughing engine that I sent
you a drawing for, after being tried for that purpose, was to
have been sent to Exeter for pumping water out of the foundations of a new bridge; but as they intend to begin their
work at the bridge before the end of March, the engine must
be there before that time, or they will erect horse machines and
not use the engine. I have therefore been obliged to send the
small boiler that I had for that purpose to Hayle Foundry, and
get the castings made there for this engine to get it in time to
prevent losing the order. I have also been obliged to take the
small portable engine from Wheal Alfred Mine and have new
apparatus fitted to it, to apply this engine for Plymouth Breakwater. A small engine, from the same patterns as Sir Charles
Hawkins' thrashing engine, which I had at work in a mine, I

have been obliged to send to one of the farmers at Padstow for thrashing, instead of one of those engines that I ordered from you. I expect that the people who ordered the engine for the West Indies are also tired of waiting. I have two other applications for engines for the West Indies, and the Messrs. Fox will want a great many engines of that size for the Plymouth Breakwater. They are to provide machinery, with every other expense, and I am to have a certain proportion of what I can save over what it now costs them to do it by manual labour. I think I have made a very good bargain, for if the plan succeeds I shall get a great deal of money, and if it fails I shall lose nothing. They have engaged with the Government to deliver 3,000,000 tons, for which they have a very good price, even if it was to be done by men's labour. I hope I shall get the engine soon on the spot, and will then let you know the result. As the boiler that was intended for the ploughing engine is to be sent to Exeter, I wish you to finish that engine with boiler, wheels, and everything complete for ploughing and thrashing, as shown in the drawing, unless you can improve on it. There is no doubt about the wheels turning around as you suppose, for when that engine in Wales travelled on the tramroad, which was very smooth, yet all the power of the engine could not slip around the wheels when the engine was chained to a post for that particular experiment.

"That new engine you saw near the seaside with me is now lifting forty millions 1 foot high with 1 bushel of coal, which is very nearly double the duty that is done by any other engine in the county. A few days since I altered a 64-inch cylinder engine at Wheal Alfred to the same plan, and I think she will do equally as much duty. I have a notice to attend a mine meeting to erect a new engine equal in power to a 63-inch cylinder single, which I hope to be able to send to you for. I have also an appointment to meet some gentlemen at Swansea, to erect two engines for them, one to lift water, the other coal, which you will hear more about, I expect, soon. If I can spare a few days when at Swansea, I will call to see you at Bridgenorth. I have not seen Mr. Richard since you left, but will call on him in a few days and do as you request. If you think the fly-wheel

is not sufficiently heavy for his engine, add half a ton more to
the ring.

"If you cannot finish all these engines at the same time, I
would rather the smaller ones should be finished first and Mr.
Richards' stand a little, because if his engine was now ready he
would not pull down his thrashing machine until he had nearly
thrashed all his corn, and the machine now stands on the spot
where the mill is to be erected.

"If I call on you from Swansea I think I shall be able to
show you a new idea, which I think will, if carried into practice,
be of immense value. Please to write to me and say particularly how you are getting on, and when you are likely to finish
the engines ordered.

"R. T."

Trevithick had sent a drawing of a *ploughing* engine
to Rastrick at Bridgenorth, that the castings might
be made, while he himself was having the boiler and
wrought-iron work constructed in Cornwall. The
engine had been ordered as a portable pumping engine,
for removing water from the foundation of a bridge at
Exeter; but before sending it to its destination, he
had arranged to plough with it, as a means of perfecting the plans and drawings for a more suitable ploughing engine then in construction, to be fitted "with
boiler, wheels, and everything complete for ploughing
and thrashing, as shown in the drawing." The *friction
of the wheels on the ground would be greater than the power
of the engine*; therefore they would not slip when the
full power was applied to draw a plough any more than
the Welsh engine, the wheels of which did not slip
through resting on smooth iron.

One of his small engines, which had been at work in
a mine, was sent as a thrashing engine to Padstow. It
is evident that, having given a portion of his attention
for a year or two to the question of steam agriculture,

he had so far progressed in 1813 as to construct thrashing machines, portable agricultural engines, and steam-
ploughs to be moved by wheels as in locomotives;
reaping, sowing, and other work, was also in future to
be the work of the steam-engine.

A drawing by Trevithick—having as usual neither
name, date, nor scale, nor writing of any kind, but the
watermark in the paper is 1813—illustrates his ideas
expressed to Sir Charles Sinclair in 1812:—"It is my
opinion that every part of agriculture might be performed by steam." The thrashing and grinding engines
were at work, and the tormenting harrowing engine
was probably designed for bringing under steam culti-
vation the extensive commons referred to. In those
days, before the practice of underground drainage, the
surface of cultivated land was thrown into furrows, or
a series of small hills and vales, the latter acting as the
surface drain for carrying off the water.

Suppose the first step in cultivating a common to be
the breaking of the soil, and throwing it into uniform
lines of rise and fall that facilitated drainage without
inconveniencing the tillage, what better machine could
have been devised than Trevithick's? A combination
of the modern tormentor and harrow loosened the
ground to the required depth, which was then, by a
revolving wheel with spades, thrown on one side, resulting in uniform lines of ridges and hollows. The
steam-shoveller was removed, or the tormentor irons
raised, when only the harrow was required.

The absence of the ordinary shafts at the front end
of the framing indicates that the spade-tormentor was
not to be drawn by horses, but whether by a locomotive or by a fixed engine is not self-evident.

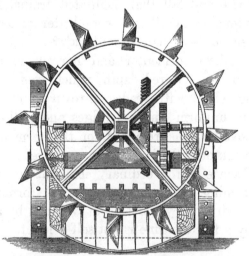

TREVITHICK'S STEAM SPADE-TORMENTOR, 1813.

[Rough draft.]

"Mr. Kendal,   "Camborne, *January 26th*, 1813.

"Sir,—I have yours of the 17th inst. The thrashing machine engine is ready for you, and shall be sent up immediately. I wish you to get about 100 fire bricks, 200 common bricks, 20 loads of stone, and 20 bushels of lime. The house will get finished while I am fixing the engine. About 1500 or 1600 weight of iron for your engine has been sent to the Blue Hills Mine, St. Agness. I wish you could send down your cart to fetch it from there to Padstow. There is no part of these castings but may be easily conveyed in a common butt or cart. When you have the stone, brick, and lime ready, and a cart to send to St. Agness for the castings, please to write me, and I will come to Padstow at the same time with them, and finish the engine. The sooner you get ready the better, as I expect to have an engagement in about four weeks' time, that would prevent my coming to Padstow for some time; therefore I wish to get your engine finished before that time. Please to write me as early as possible, and let me know when you will be ready for me, and what day I shall meet your cart at St. Agness for the castings.

"Your obedient servant,

"Richard Trevithick."

Real inventors hesitate not to erect their own engines, lend a hand in building the house, walk to the scene of action, or take a lift in a cart; and by such steps was the gift of genius moulded to the wants of daily life; while the modern engineer of eminence, living in large cities, knows little of the minutiæ of his work, or even of the working mechanics on whose skill the success of his ideas is dependent.

"In 1815, Mr. Kendal, the proctor of Padstow, sent for me to repair his steam-engine. To prevent the old disputes in collecting his corn tithes, he had at work one of Captain Trevithick's steam thrashing machines. The small farmers sent their corn produce to him to be thrashed; the grain was measured, the tenth taken out, the remainder returned to the farmer. The three-way cock, which worked the engine, was joined in its shell; on freeing it the engine continued to work very well." [1]

" In 1818 I put a new four-way cock to a thrashing engine that Captain Trevithick had made for Mr. Kendal, of Padstow, who was the receiver of tithe corn. The boiler was a tube of wrought iron, about 4 feet in diameter and 6 feet long, standing on its end. The cylinder was fixed in the top of the boiler; an upright from the top of the cylinder supported the fly-wheel shaft; a connecting rod from the crank-pin in the fly-wheel was fastened to a joint-pin in the piston. The cylinder had no cover. The four-way cock was worked by an excentric on the shaft, moving a lever, which was kept in contact with the excentric by a spring." [2]

" About 1824 I worked in Binner Downs Mine one of Captain Trevithick's puffer whim-engines. The boiler was cylindrical, made of wrought iron. It stood on its end, with the fire under it, and brick flues around it. The cylinder was let down into the top of the boiler. A four-way cock near the top of the cylinder turned the steam on and off. The fly-wheel and its shaft were fixed just over the cylinder. A lever and rod worked the four-way cock and feed-pole. The waste steam puffed through a launder into the feed-cistern. The cylinder was about 12 inches in diameter, with a 3-feet stroke." [3]

Mr. Kendal's steam thrashing machine remained at work at least six years, during which time the only apparent repair was the four-way cock, worked by an excentric, which, if neglected, was apt to stick fast in its shell.

One of the puffer-whims erected about this time was similar to the thrashing engine for Padstow, differing from the earlier one made for Sir C. Hawkins, in

[1] Captain Samuel Grote's recollections, 1858.
[2] Recollections of Captain H. A. Artha, Penzance, 1868.
[3] Recollections of Henry Vivian, Harvey and Co.'s Works, 1869.

having a portable boiler so arranged that if necessary it could be easily placed on wheels.

[Rough draft.]

" SIR, "CAMBORNE, *March 15th*, 1813.

" I have your favour of the 11th inst. respecting a steam-engine for thrashing. I have made several, all of which answer the purpose exceedingly well. They are made on a very simple construction so as to be free from repairs, and are kept in order and worked by the farm labourers, who never before saw a steam-engine. The first I made on this plan was for Sir Christopher Hawkins, who resides at this time in Argyll Street, Oxford Street, London. If you call on him, he, I doubt not, would give you every information you require respecting its performance. This was a fixed engine, because it was only required to work on one farm. It has been at work nearly eighteen months, and has not cost anything in repairs, nor any assistance but from the labourer who puts in the corn; he only gives three or four minutes every hour to put on a little coal. A few pails of water, put into the furnace in the morning, is sufficient for a day's work. They have at different times tried what duty the engine would perform with a given quantity of coal, and found that two Cornish bushels, weighing 168 lbs., would get up steam and thrash 1500 sheaves of wheat in about six hours.

" Before this engine was erected, they usually thrashed 500 sheaves, with three heavy cart-horses for a day's work. I cannot say exactly the measurement of the corn that it thrashed, but it was considerably above 60 winchesters of wheat with 168 lbs. of coal; not a halfpenny in coal for each winchester of wheat.

" The engines that I have since erected have performed the same duty.

" The horse machinery is thrown out of use, but the same drum is turned by the engine.

" A fixed engine of this power I would deliver to you in London for 100 guineas; it would cost you about 15*l.* more to fix the furnace in brickwork.

" A portable engine costs 160 guineas, but it would cost nothing in erecting, as it will be sent with chimney and every thing complete on its own wheels (the drum, &c., excepted), which you may convey with one horse from farm to farm as easy as a common cart.

" If you have not sufficient work for it you can lend it to your neighbours. The last engine I erected was about three weeks since, for a farmer that kept four horses and two drivers. The parts of the horse machine thrown out of use, together with the four horses, sold for more money than he gave me for the engine, exclusive of 4*l.* per week that it cost him in horse keep and drivers to thrash 3000 sheaves per week.

" Now the engine performs more than double that work, and does not cost above 10*s.* per week; and the labourer in the barn does double the work he did before for the same money. If you wish the same engine to have sufficient power to turn one pair of mill-stones, the cost will be 220 guineas.

" R. TREVITHICK.

" MR. J. RAWLINGS, *Strood, Kent.*"

" CAMBORNE, 28*th August*, 1813.

" MESSRS. HAZELDINE, RASTRICK, AND CO.,
    "Gentlemen, — Lord Dedunstanville's engine thrashed yesterday 1500 sheaves in 90 minutes with 40 lbs. of coal.

" RD. TREVITHICK."

The first steam thrashing engine was worked by a labouring man for eighteen months, without needing repair, or even attention beyond three or four minutes each hour to put on a little coal.

Necessary stoppages for various purposes caused a day's work to be no more than the engine could perform in half a day. No additional feed-water was required during an ordinary day's work to thrash 1500 sheaves of wheat with 168 lbs. of coal, while on a special occasion that quantity was thrashed in an hour and a half, consuming only 40 lbs. of coal. Three horses during three days were required to do the same amount of work. A farmer sold his horses used in thrashing for more money than his engine cost, which did twice as much work at a reduced expenditure, and also saved the feed of the horses.

Such an engine could be delivered in London for 100 guineas, while a portable engine on wheels with a differently constructed boiler, requiring no mason work, would cost 160 guineas.

[Rough draft.]

" GENTLEMEN,   " CORNWALL, CAMBORNE, 19th *August*, 1813.

" I have your favour of the 9th August, respecting steam-engines for St. Kitts. I fear it will not be possible to get an engine ready by the 1st of November.

" As you say the gentleman that is about to take them out is a clever man, and likely to promote the use of them, I will make immediate inquiry, and, if possible, will get one ready, of which I will inform you in a short time.

" I very much wish that every person who intends to employ a steam-engine of mine would first examine the engine, and be satisfied with the construction before giving an order, for which reason I must request you to send your friend down to Messrs. Hazeldine and Rastrick's foundry, Bridgenorth, Shropshire, where he may see the portable steam-engine that was made for Mr. Pickwood, which the founders will set to work for his inspection in half an hour after his arrival. As this gentleman has a taste for machines, and wishes to make himself fully acquainted with the principle and use of the steam-engine, he will be much gratified with the sight of this curious machine and with the information he will receive from the founders, which will be essentially necessary to him before leaving England.

" I am extremely disappointed that this engine was not forwarded to Mr. Pickwood, as I find from his letter that he has an exceedingly clever and active mind, and is a very fit person to take the management of introducing a machine into a new country.

" This engine is engaged by a Spanish gentleman, who is

going to take out nine of my engines with him to Lima, in South America, in about six weeks.

"I remain, your obedient servant,

"RICHD. TREVITHICK.

"MESSRS. PLUMMER, BARHAM, AND CO.,
   "London.

"N.B.—If your friend goes to Bridgenorth, let him show this letter to the founders."

The engine, intended for the West Indies, so pleased Mr. Uville, that he begged to have it made over to him for South America, where it worked the machinery for rolling gold and silver in the Mint at Lima.

"About 1815, while erecting a high-pressure pole-engine at Legassack for Mr. Trevithick, and doing some repairs to Mr. Kendal's thrashing engine, a Creole, I think called Nash, brought a note from Captain Trevithick, stating that the bearer was anxious to be taught to erect and work the portable engines for Jamaica.

"Sir Rose Price, who had property in the West Indies, had sent him to Mr. Trevithick for that purpose." [1]

It is therefore probable that some of Trevithick's engines reached Jamaica. Sir Rose Price was well known to Lord Dedunstanville and Sir Charles Hawkins, and living near them, saw the engines at work and their fitness for his property in Jamaica.

Lord Dedunstanville's engine of 1812 was sold as old iron to Messrs. Harvey and Co. not long before 1843. Having remained for some time on the old-scrap heap, it was in that year again worked to drive machinery. Instead of the original rope-driver on the fly-wheel, a chain was used, the links of which caught on projecting pins on the driving wheel. In that form it continued to work until 1853, before which it was fre-

---

[1] Recollections of Captain H. A. Artha, Penzance, 1869.

quently seen by the writer prior to its removal to make room for a more powerful engine.

What greater proof could be given of the fitness of design of this early engine, than its long life of forty years under such rough treatment, and the facility with which it was applied to different uses. Mr. Bickle, who, from recollection, had made a sketch of this engine before the writer had found Trevithick's sketch, says that after the engine had ceased to work, the boiler was turned to account in heating tar in the ship-builder's yard.

"In 1854 I saw working in a shed at Carnsew, in the ship-building yard of Harvey and Co., of Hayle, an engine working a stamps for pounding up the slag and furnace bottoms from the brass-casting foundry.

"I was then the foreman hammerman in Harvey and Co.'s smiths' shop and hammer-mill, and frequently noticed this old engine and inquired about it. It had been brought from Lord Dedunstanville's, at Tehidy Park, where it at one time worked a thrashing machine. The boiler was of wrought iron, built in brickwork, and looked like a big kitchen-boiler. A flattish cover was bolted on to the top of the boiler, and the cylinder was let down into this top.

"The cylinder had no cover; it was about 8 or 10 inches in diameter and 2 or 3 feet stroke. The piston was a very deep one, with a joint for the connecting rod which went direct to the crank, which was supported on two upright stands from the cover on the boiler. The fly-wheel had a balance-weight for the down-stroke. A pitch-chain for driving passed over the wheel, which had pins in it, or projections, to catch into the square links of the driving chain; it was worked by a four-way cock."[1]

"About 1843, when we were building iron boats for the Rhine, the old engine was put to work to drive the tools or machinery in the yard. She was very useful to us and worked very well. She worked about ten years, and was then thrown

---

[1] Recollections of Banfield, foreman with Harvey and Co., Penzance, 1869.

out to make room for a new and larger engine for our saw-mills.
The chain-wheel for driving was made here, it did not belong
to it originally." [1]

"My father (then the foreman boiler-maker) about twenty-
four years ago took the old engine from the scrap heap, where
it had been for many years, and set it to work in the tool shop.
My father said it had come from Tehidy as old iron." [2]

The use of the high-pressure steam agricultural
engine was not confined to Cornwall. Mr. H. Pape,
still carrying on business in Hazeldine and Rastrick's
old engine manufactory at Bridgenorth, says:—

"My father worked as a smith under Mr. Rastrick. Mr.
Hazeldine had the foundry when Trevithick's engines were
made, and have heard my father speak of them. I have seen
three of them at work in Bridgenorth; one of them at
Mr. Jasper's flour-mill, it drove four stones, and continued in
work up to 1837; one at Sing's tan-yard worked up to 1840;
and one was on Mr. Jasper's farm at Stapleford for doing farm
work. Mr. Smith, now on the farm, worked it up to about 1858.

"The engines that worked in Bridgenorth had cast-iron
cylinders for the outer casing of the boiler, one cylinder for
small engines, three or four cylinders bolted together for the
larger ones. The fire-tube was wrought iron, the chimney stood
up by the fire-door. The cylinder was let down into the boiler;
it worked with a four-way cock. There was a piston-rod, cross-
head, two guide-rods on the top of the cylinder, and two side
rods to the crank and pin in 'the fly-wheel.'" [3]

"My first husband had to do with the foundry; his father,
Mr. Hazeldine, was a partner with Mr. Davies and Co. in 1816.
In 1817 the partnership was broken up, and the foundry carried on by Hazeldine. I used to have two or three drawers full
of drawings and account-books that were brought from the

[1] Recollections of Mr. Warren, master ship-builder, Harvey and Co., 1869.
[2] Recollections of Mr. Burral, jun., master boiler-maker, Harvey and Co.,
1869.
[3] Recollections of Mr. William H. Pape, Bridgenorth, 13th June, 1869.

works. I kept them for many years, but now the greater part of them have gone to light the fire ; all the drawings are gone."[1]

The engines described by Mr. Pape are of the type made by Trevithick, in Wales, about 1804, having a fire-place in the boiler, and similar in form to the Welsh locomotive.

The drawings which served to light the fires certainly included Trevithick's plans for the steam-locomotive, ploughing engine, the screw-propeller, and many others of equal interest.

"DEAR SIR,   "STABLEFORD, *March 26th*, 1870.

"My grandfather's name was John Jasper, Esq., of Stableford; he must have been one, if not the first, user of a steam-engine for thrashing, winnowing, and shaking the straw all at one operation ; it may have been erected eighty years ago, for an old servant of the family just now dead, aged ninety, worked when a boy in the steam-mill at Bridgenorth erected by my grandfather about the same time.

"The thrashing engine was a side-lever engine, worked with a three side-way cock and tappet, a cylinder about 8¼ inches in diameter, and a 3 feet 4 inch stroke, cast-iron crank-shaft, cross-head, and guides. The boiler was placed underneath the engine, the fire under it, with brick flues. The boiler was about 9 feet long and 4 feet diameter.

"The old side rods made of wood are still here, and so was the engine until about twelve years ago. I sent the cylinder, &c., to Coalbrookdale.

"I am, Sir,

"Yours truly,

"THOMAS SMITH."

The Stableford agricultural engine was probably made in 1804. The cylinder, of 8¼ inches in diameter, is precisely the size of that in the Welsh locomotive,

---

[1] Recollections of Mrs. Marm, Bridgenorth, 12th June, 1869.

but the stroke was reduced from 4 feet 6 inches to 3 feet 4 inches, being very nearly the same as the Newcastle locomotive. The cross-head, side rods, and boiler were very similar to the Welsh stationary engines of that date. This engine remained in use more than fifty years.

The engines specially referred to in this chapter fully prove, from their length of service, the practical character of Trevithick's inventions, and of his having persevered with his high-pressure portables until their usefulness as locomotives and as agricultural helps had been established; but the ploughing, though fully designed, and probably put into practice, was not followed up to the same approach to perfection, or the record of its progress has been lost.

Since the foregoing was written, the following has been received :—

"DEAR SIR,   "TREWITHEN, PROBUS, *May* 17*th*, 1872.

"The engine you refer to is still occasionally used here; when first erected there was a large quantity of corn thrashed by it, but of late years it has not been much used except for chaffing, bruising, &c.

"I remain, dear Sir,
"Yours truly,
"WM. TRETHNOY.

"F. TREVITHICK, Esq."

Trevithick's Trewithen engine, which sixty years ago was more manageable than horses going momentarily faster or slower at the will of a common labourer,[1] remains in use unchanged.

His preparations for South America, and application of high steam in the large Cornish pumping engines, interfered with the perfecting the smaller agricultural work.

---

[1] See vol. ii., p. 38.

CHAPTER XIX.

WHEN in the autumn of 1810 Trevithick returned to Cornwall, the experience of ten busy years had established the practicability and usefulness of the high-pressure engine. The principles of the invention were now to be applied on engines of the largest size.

In 1811, the late Captain S. Grose, a young pupil of Trevithick's, was employed to erect at Wheal Prosper Mine, in Gwythian, the first high-pressure steam pole condensing engine. It was placed immediately over the shaft and pump-rods, requiring no engine-beam. The air-pump, feed-pump, and plug-rod were worked from the balance-bob. The pole was 16 inches in diameter, with a stroke of 8 feet. The boilers were two wrought-iron tubes, 3 feet in diameter and 40 feet long. The fire was external. Shortly after Captain H. A. Artha erected several of those pole-engines for Trevithick. The drawing shows the simplicity of parts of this highly expansive steam-engine, beginning the up-stroke with steam of 100 lbs. to the inch above the atmospheric pressure, expanding it during the stroke down to a pressure of 10 lbs., and then condensing to form a vacuum for the down-stroke. It cost 750 guineas.

The drawings of this expansive pole condensing engine are from the dimensions given by Captain Grose who erected it, and by Captain Artha who knew it well.

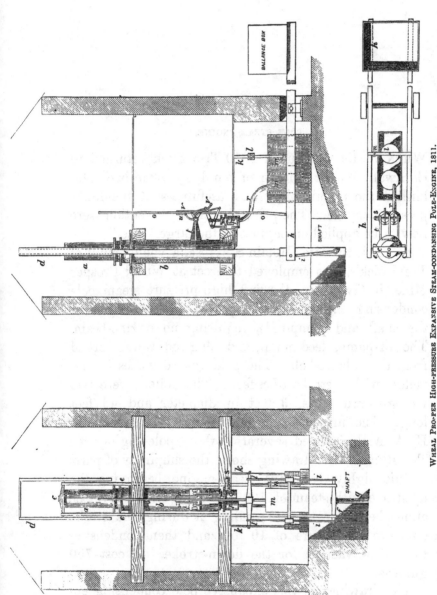

WHEAL PRO-PER HIGH-PRESSURE EXPANSIVE STEAM-CONDENSING POLE-ENGINE, 1811.

a, cast-iron pole, 16 inches diameter, 8-feet stroke; *b*, pole-case, a small bit larger in internal diameter than the pole; *c*, cross-head, fixed on top of pole; *d*, guides for cross-head; *e*, side rods connecting the two cross-heads; *f*, bottom cross-head; *g*, pump-rod; *h*, balance-beam, with box for weights; *i*, connecting rods from balance-beam to bottom cross-head; *k*, cross-head and side rods for working air-pump; *m*, air-pump, condenser, &c.; *n*, guides for air-pump cross-head; *l*, guides for cross-head; *k*, cross-head and ... *g*, steam-valve; *r*, exhaust-

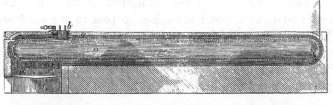

TREVITHICK'S CYLINDRICAL BOILER FOR WHEAL PROSPER ENGINE, 1811.

Detail of Boilers :—a, two wrought-iron boilers, 3 feet in diameter, 40 feet long, using steam of
100 lbs. on the square inch above the atmosphere; b, cast-iron manhole door and safety-valve;
c, ash-pit; d, fire-place; e, flues, the fire going first the whole length under the bottom of the boiler,
then back again over the top, and into the chimney; f, brickwork; g, ashes or other convenient non-
conductor of heat; the fire-place ends of the boilers were 15 inches lower than the opposite ends,
increasing the safety, with less liability to prime, and greater surface for superheating.

"When a boy I was placed as an apprentice or learner with
Captain Trevithick, before he left Cornwall for London. On
his return to Cornwall, about 1810, he employed me to erect
his first high-pressure expansive pole pumping engine at a mine
in Gwythian.

"The pole was 16 inches in diameter; the stroke was very
long, but I do not exactly recollect the length. It had a condenser and air-pump. There were two boilers made of wrought
iron, 8 feet in diameter and 40 feet long. The fire was placed
under them at one end, and flues went round them. A feed-
pump forced water into the boilers; each had a safety-valve
with a lever and weight. The steam in the boiler was 100 lbs.
to the square inch. The pole was raised by the admission of
the strong steam under its bottom. The steam-valve was closed
at an early part of the stroke, and the steam allowed to expand;

at the end of the stroke it was reduced to 20 lbs. or less, when the exhaust-valve allowed the steam to pass to the condenser, and the pole made its down-stroke in vacuum. A balance-bob regulated the movement of the engine.

"Trevithick's character in those days was, that he always began some new thing before he had finished the old."[1]

Captain Artha, one of his assistants, said :—

"I erected several of Captain Trevithick's pole-engines. My brother Richard worked the one at Wheal Prosper when first erected. The pole made an 8-feet stroke. The case was fixed over the engine-shaft on two beams of timber from wall to wall. A cross-head was bolted to the top of the pole, and from it two side rods descended to a cross-piece under the pole-case, from which the pump-rod went into the shaft. A connecting rod worked a balance-beam, which worked the air-pump, feed-pump, and plug-rod for moving the valves. The steam, of a very high pressure, worked expansively."[2]

The first admission of the high-pressure steam under the pole was equal to a force of 8 or 9 tons, causing it and its attached pump-rods to take a rapid upward spring. Having travelled 1 or 2 feet of its stroke of 8 feet, the further supply of steam from the boiler was cut off, and its expansion, together with the momentum of the mass of pump-rods, completed the upward stroke. The pressure of the steam in the pole-case at the finish of the up-stroke would be reduced to say 10 or 20 lbs. to the inch, according to the amount of work on the engine. The steam then passed to the condenser and air-pump, and the engine made its down-stroke by the vacuum under the pole, and by the weight of the descending pole and pump-rods.

Each boiler was a wrought-iron tube 3 feet in diameter and 40 feet long, the fire-place under one end,

---

[1] Captain Samuel Grose's recollections. 1858. Gwinear.
[2] Captain H. A. Artha's recollections. Penzance, 1869.

with brick flues carrying the heated air under the whole length of the bottom of the boiler, and back again over the top or steam portion for superheating.

[Rough draft.]

"CAMBORNE, 28*th* *February*, 1813.

"I will engage to erect a puffer steam-engine, everything complete at the surface, on the Cost-all-lost Mine, capable of lifting an 8-inch bucket, 4½-feet stroke, twenty-four strokes per minute, 30 fathoms deep, or 280 gallons of water per minute from that same depth, being a duty equal thereto, for 550 guineas. But if a condensing engine, 600 guineas. If of the same size as Wheal Prosper, 750 guineas.

"RICHARD TREVITHICK."

The engines, erected in 1811 or 1812, combined the novelty of the steam pole-engine, with the use of high-pressure steam of 100 lbs. on the square inch, and the comparatively untried principle of steam expansion, carried to what in the present day is thought an extreme and unmanageable limit.

The Wheal Prosper engine fixed near the sea-shore at Gwythian is referred to in Trevithick's note to Mr. Rastrick,[1] as "that new engine you saw near the seaside, with me, is now lifting forty millions, 1 foot high, with 1 bushel of coal" (84 lbs.), "which is very nearly double the duty that is done by any other engine in the county."

This was probably the first application of high-pressure steam to give motion to pump-rods. The engine, as compared with the neighbouring Watt low-pressure steam vacuum pumping engines, was small, but the principles of high steam, expansive working, and va-

---

[1] See Trevithick's letter, January 26th, 1813, chap. xviii.

cuum, were combined successfully to an extent scarcely ventured on by modern engineers.

Trevithick's high-pressure condensing whim-engines had been for some years at work in Cornwall, but mine adventurers had not dared to risk the application of high-pressure steam to the large pumping engines, fearing its great power would prove unmanageable, and its rapid movement cause breakage of the pump-rods and valves.

Two distinct inventions or improvements, each of which was actually followed up in different mines, show themselves in this engine: one being the form of boiler to give with economy and safety high-pressure expansive steam for large engines; the other, the application of a pole in lieu of a piston, as a more simple engine for working with strong expansive steam, and more easily constructed by inexperienced mechanics, who had none of the slide lathes or planing machines so much used by engine builders of the present day.

"About 1814 Captain Trevithick erected a large high-pressure steam-puffer pumping engine at the Herland Mine. The pole was about 30 inches in diameter, and 10 or 12 feet stroke. There was a cross-head on the top of the pole, and side rods to a cross-head under the pole-case. The side rods worked in guides. The pole-case was fixed to strong beams immediately over the pump-shaft. The steam was turned on and off by a four-way cock. The pressure was 150 lbs. to the inch above the atmosphere. The boilers were of wrought iron, cylindrical, about 5 feet 6 inches in diameter and 40 feet long, with an internal tube 3 ft. in diameter. The fire-place was in the tube. The return draught passed through external brick flues."[1]

"When a young man, living on a farm at Gurlyn, I was sent to Gwinear to bring home six or seven bullocks. Herland Mine

---

[1] Recollections of the late Captain Charles Thomas, manager of Dolcoath Mine.

was not much out of my way, so I drove the bullocks across Herland Common toward the engine-house. Just as the bullocks came near the engine-house the engine was put to work. The steam roared like thunder through an underground pipe about 50 feet long, and then went off like a gun every stroke of the engine. The bullocks galloped off—some one way and some another. I went into the engine-house. The engine was a great pole about 3 feet in diameter and 12 feet long. A cast-iron cross-head was bolted to the top of the pole. It had side rods and guides. A piece of iron sticking out from the cross-head carried the plug-rod for working the gear-handles. The top of the pole worked in a stuffing box. A large balance-beam was attached to the pump-rods, near the bottom cross-head.

"There were two or three of Captain Trevithick's boilers with a tube through them, the fire in the tube. They seemed to be placed in a pit in the ground. The brick flues and top of the bricks were covered with ashes just level with the ground. A great cloud of steam came from the covering of ashes.

"I should think the pressure was more than 100 lbs. to the inch. People used to say that she forked the mine better than two of Boulton and Watt's 80-inch cylinder engines. We could hear the puffer blowing at Gurlyn, five or six miles from the Herland Mine.

"In 1813 I carried rivets to make Captain Trevithick's boilers in the Mellinear Mine; they were 5 feet in diameter and 30 or 40 feet long, with an internal fire-tube. It took four or five months to build them. In the present day (1869) a fortnight would build them. The largest boiler-plates obtainable were 3 feet by 1 foot. We had to hammer them into the proper curve. The rivet-holes were not opposite one another. A light hammer was held against the rivet-head in riveting, in place of the present heavy one, so the rivet used to slip about, and the plates were never hammered home so as to make a tight joint."[1]

Lest the reader should doubt the comparative power of the Watt low-pressure vacuum and Trevithick's high-

---

[1] Recollections of Mr. James Banfield, Penzance, 1871.

pressure steam-engines, a short but sufficiently close calculation shows that taking Stuart's[1] estimate of the effective power of the Watt engine at 8½ lbs. on each square inch of the piston, and Trevithick's engine at anything approaching to 150 lbs. on each square inch, it becomes evident that the latter would be ten or twenty times more powerful than the former. A few figures will put the question in more practical form.

The Wheal Prosper 16-inch pole high-pressure expansive steam vacuum engine commenced its up-stroke with steam of 100 lbs. on the inch, acting on the 122 square inches of the pole, which steam at the finish of the stroke was reduced by expansion to 10 lbs., giving, say, an average steam pressure of 55 lbs. The down-stroke was caused by a vacuum under the pole of 14 lbs. on the inch, reduced by, say, one-third loss in working the air-pump to 9 lbs., giving from the compound stroke a force of 64 lbs. on each square inch, which, multiplied by the area of the pole, gives a net force of 7808 lbs.

The Herland 33-inch pole high-pressure expansive steam puffer-engine commenced its up-stroke with steam of 150 lbs. on the inch, acting on the 855 square inches of the pole, which steam at the finish of the stroke—we will suppose—was reduced by expansion to 75 lbs., giving an average steam pressure of, say, 112. As this puffer-engine used no vacuum, the down-stroke gave no increase of power; its compound stroke was therefore a force of 112 lbs. on each square inch, which, multiplied by the area of the pole, gives a net force of 95,760 lbs.

To compare the Trevithick high-pressure steam pumping engine, with the Watt low-pressure steam

---

[1] See Stuart's 'History of the Steam-Engine.'

pumping engine, take one of the largest of the latter, made about that time, say, with an 80-inch cylinder, which commenced its down-stroke with steam of, say, 3 lbs. on the inch, acting on the 5000 square inches of the piston, which steam at the finish of the stroke —the writer is describing the usage at that time, for Watt himself advocated a less steam pressure— was reduced by expansion to, say, 1 lb., giving an average steam pressure of, say, 2 lbs. on the top of the piston, whose under side was in vacuum equal to 14 lbs. on the inch, reduced by, say, one-third loss in working the air-pump to 9 lbs., which power, from vacuum added to the 2 lbs. from steam, gives a net force of 11 lbs. on each square inch of the piston. As the Watt pumping engine moved in equilibrium during its up-stroke, it thereby gained no increase of power; its compound stroke was therefore a force of 11 lbs. on each square inch, which, multiplied by the area of the piston, gives a net force of 55,000 lbs.

The practical comparison therefore stands,—Trevithick's 16-inch pole high-pressure steam, and vacuum, on each inch 64 lbs., net force 7808 lbs.; Trevithick's 33-inch pole high-pressure steam, without vacuum, on each inch 112 lbs., net force 95,760 lbs.; Watt's 80-inch piston, low-pressure steam, and vacuum, on each inch 11 lbs., net force 55,000 lbs. As the first cost was mainly dependent on the size, the Trevithick engine was commercially much more valuable than the Watt engine.

"I saw Captain Trevithick's puffer working at the Herland Mine. The steam used to blow off like blue fire—it was so strong. The lever on the safety-valve was about 3 feet long, with a great weight on it, more than a hundredweight. The engine did not answer very well, for the packing in the pole stuffing box used to burn out, and a cloud of steam escaped. The greatest difficulty was in the leaking of the boilers. You could hardly go near them. Before that time we always put rope-yarn between the lap of the boiler-plates to make the seams tight. Captain Dick's high-pressure steam burnt it all out. He said, 'Now you shall never make another boiler for me with rope-yarn.' Everybody said it was impossible to make a tight boiler without it. We put barrowfuls of horse-dung and bran in Captain Dick's boilers to stop the leaks."[1]

This difficulty of making a tight and safe boiler, that puzzled Watt, was moonshine to Trevithick. When the strained boiler and flinching rivets allowed the boiler-house to become full of dense steam, Trevithick told them to cover it up with ashes, they would not see it quite so much then, and it would keep the heat in the boiler. Bran or horse-dung inside was a good thing as a stop-gap, though it added not to the strength of the boiler. Trevithick was himself in a cloud of steam in the engine-house; yet, with such surroundings, he turned on and off his gunpowder steam, from his cannon of a pole-case, of 40 tons force, sending his bolt-shot pole, 33 inches in diameter, its destined course of ten feet, and back again, as though it were a shuttlecock, several times in a minute.

Having by one or two years of experience proved the value of his new pole-engine, he applied for a patent on the 13th June, 1815,[2] of which the following is the portion referring particularly to the pole-engine :—

"Instead of a piston working in the main cylinder of the steam-engine, I do use a plunger-pole similar to those employed in pumps for lifting water, and I do make the said plunger-pole nearly of the same diameter as the working cylinder, having

---

[1] Henry Clark of Redruth, in 1869, aged eighty-three years.
[2] See full copy of patent, chap. xvi.

only space enough between the pole and the cylinder to prevent friction, or, in case the steam is admitted near the stuffing box, I leave sufficient room for the steam to pass to the bottom of the cylinder, and I do make at the upper end of the cylinder for the plunger-pole to pass through a stuffing box of much greater depth than usual, into which stuffing box I do introduce enough of the usual packing to fill it one-third high. Upon this packing I place a ring of metal, occupying about another third part of the depth of the stuffing box, this ring having a circular groove at the inside, and a hole or holes through it communicating with the outside, and with a hole through the side of the stuffing box; or, instead of one ring containing a groove, I sometimes place two thinner rings, kept asunder by a number of pillars to about the distance of one-third of the depth of the stuffing box, and I pack the remaining space above the ring or rings, and secure the whole down in the usual manner. The intention of this arrangement is to produce the effect of two stuffing boxes, allowing a space between the two stuffings for water to pass freely in from the boiler or forcing pump through a pipe and through the hole in the side of the stuffing box, so as to surround the plunger-pole and form the ring of water for the purpose of preventing the escape of steam by keeping up an equilibrium between the water above the lower stuffing and the steam in the cylinder. By this part of my said invention I obviate the necessity of that tight packing which is requisite when steam of a high pressure is used, and consequently I avoid a greater proportion of the usual friction, because a very moderate degree of tightness in the packing is quite sufficient to prevent the passage of any injurious quantity of so dense a fluid as water. And I do further declare that I use the plunger-pole, working in a cylinder and through a double stuffing, either with or without a condenser, according to the nature of the work which the steam-engine is to perform."

Though Trevithick has been spoken of as a visionary, intractable schemer, observation shows that he adhered with tenacity to original ideas, proved to be good. The plunger-pole pump, the water-pressure engine, the

Camborne locomotive, the pole steam-engine, were all built on the same groundwork originally started with, of greatest simplicity of form, and absence of many pieces; and it may be observed that he never applied for a patent until the value of the idea had been proved by experiment.

In practice the difficulty of keeping the pole-packing in order was one of the objections to the plan; for it either leaked, or, if packed tight, caused much friction and wearing away of the middle of the pole faster than the ends, from the greater speed at the middle of the stroke. The steam-ring was therefore of importance in the engine, in those days of inaccurate workmanship; like the water cup on the gland of the plunger-pump packing, it prevented external air from injuring the vacuum.

"MR. GIDDY,   "CAMBORNE, *July 8th*, 1815.

"Sir,—About a fortnight since I received letters from Lima, and also letters to the friends of the men who sailed with the engines. They arrived on the 29th January, after a very good passage, and without one hour's sickness. Both their and my agreements were immediately ratified, and they are in high spirits. The ship finished discharging on the 11th February, which was the day those letters sailed from Lima with $12,000 for me, which has all arrived safe.

"I shall make another fit-out for them immediately. I expect that all the engines will be at work before the end of October; half of them must be at work before this time. The next day, after their letters sailed for Europe, they intended to go back to the mines. Woolf's engine is stopped at Herland, and I have orders to proceed. A great part of the work is finished for them, and will be at work within two months from the time I began. I only engage that the engine shall be equal to a B. and Watt's 72-inch single, but it will be equal to a double 72-inch cylinder. It is a cast-iron plunger-pole, over the shaft, of 33 inches diameter, 10-feet stroke. The boiler is two

tubes, 45 feet long each, 3 feet diameter, ½ an inch thick, of
wrought iron, side by side, nearly horizontal, only 15 inches
higher at the steam end of the tubes, to allow the free passage
of steam to the steam-pipe. There are two 4-inch valves, one the
steam-valve, the other the discharging valve. I have made the
plunger-case and steam-vessel of wrought iron ¾ of an inch thick.
The steam-vessel is 48 inches in diameter. The plunger stands
on beams over the shaft, with the top of it at the level of the
surface, with a short T-piece above the plunger-pole, and a side

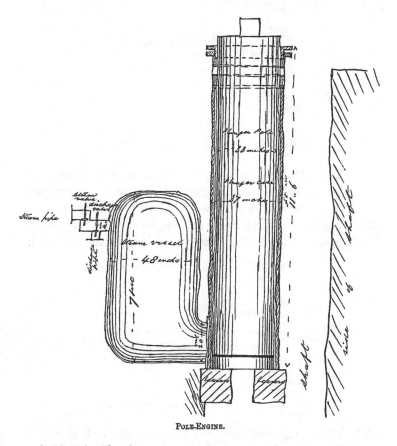

POLE-ENGINE.

rod on each side, that comes up between the two plunger-beams
in the shaft; this does away with the use of an engine-beam,
and the plungers do away with the use of a balance-beam.

"The fire is under the two tubes, and goes under them for 45 feet, and then returns again over them, and then up the chimney. Those tubes need no boiler-house, because they are arched over with brick, which keeps them from the weather, and scarcely any engine-house is needed, only just to cover the engineman.

"Suppose a 72-inch cylinder (having 4000 inches), at 10 lbs. to the inch, an 8-feet stroke, working nine strokes per minute (which is more strokes of that length than she will make when loaded to 10 lbs. to the inch).

Inches.
4000 in a 72-inch cylinder, single.
  10 lbs. to the inch.
  ———
40000
    8 feet stroke.
  ———
320000
    9 strokes per minute.
  ———
2880000 lbs. lifted one foot high per minute.

"Suppose a 33-inch plunger-pole, 10-feet stroke, ten strokes per minute (which is not so fast by three or four strokes per minute as this engine will go, because she will have no heavy beam to return, neither will she have to wait for condensing, like B. and Watt's, which, when loaded, hangs very long on the injection).

855 square inches in a 33-inch plunger-pole.
  10 strokes per minute.
  ———
8550
  10 feet stroke.
  ———
85500
    34 lbs. to the inch, real duty.
  ———
342000
256500
  ———
2907000 lbs. lifted one foot high per minute.

"I should judge that less than 50 lbs. to the inch above the atmosphere would be quite enough to do the work of a 72-inch cylinder single, which is but a trifle for those wrought-iron tubes to stand. This engine, everything new, house included, ready

for work, will not exceed 700*l.* Two months are sufficient for
erecting it. The engine of Woolf's, at Wheal Vor, which is but
two-thirds the power of a 72-inch cylinder, single power, cost
8000*l.*, and was two years erecting. I would be much obliged
to you for your opinion on this business.

"I remain, Sir,

"Your very humble servant,

"RICHARD TREVITHICK.

"I am sorry to say that the mines in general are very
poor."

He shows that with steam of 34 lbs. to the inch, his
Herland pole puffer steam-engine of 33 inches in diameter would be equal in power to the Watt low-
pressure steam vacuum engine, with a 72-inch cylinder.
Herland, like Dolcoath and Wheal Treasury, was the
chosen battle-ground of rival engineers; fifty years
after Newcomen had there erected his famously large
70-inch cylinder engine, Watt surpassed it in size by
a cylinder 2 inches more in diameter, and, after personally superintending its erection in 1798, declared
that "it could not be improved on." Mr. Davey, the
mine manager, considered that it did twenty millions of
duty, though Mr. Watt had made it twenty-seven
millions with a bushel of coal.[1] This difference is probably explained by the then Cornish bushel weighing
84 lbs., while Watt generally calculated a bushel at
112 lbs.

Trevithick declining to believe Watt's prognostica-
tion, a public test of Watt's engines in the county
was demanded; Mr. Davies Gilbert, with Mr. Jenkin,
were requested to report on their duty, and gave it in

---

[1] Lean's 'Historical Statement of the Steam Engine, p. 7.

1798 as averaging seventeen millions.[1] During the same year the adventurers in Herland Mine engaged Trevithick and Bull, jun., to erect a 60-inch cylinder Bull engine to compete with Watt's 72-inch cylinder, The result of this fight is not traceable, nor what took place there during the succeeding fifteen years; when in 1814 Woolf erected in Cornwall his double-cylinder engine to compete with Watt's engine, and Trevithick attacked them both with his Herland high-pressure pole puffer in 1815, when he erected at his own risk and cost a 33-inch pole-engine, engaging that it should, both in power and economical duty, equal the Watt 72-inch engine. The boilers were similar in form to those used a year before in Wheal Prosper high-pressure steam vacuum pole-engine, being two wrought-iron tubes, each 45 feet long and 3 feet in diameter, made of plates half an inch thick. The fire was in external flues. The engine was fixed directly over the pump-rods in the shaft, using neither main beam nor air-pump.

Trevithick's rough hand-sketch shows the steam-ring in the stuffing box and the steam-vessel; the particular use of the latter he has not described : probably it was because Cornish pumping engines, not having the controlling crank to limit the movement of the piston, are obliged to trust to the very admirable, but little understood, steam-cushion, without which the ascending piston would inevitably strike and break the cylinder-cover, while in the pole puffer-engine this danger was during the descent of the pole, and therefore the discharge-steam valve was closed, while the steam in the pole-case was still of ten or more pounds to the inch, so that by the time the pole reached the finish of its down-stroke, it had compressed this steam-cushion,

---

[1] Lean's ' Historical Statement of the Steam-Engine,' p. 7.

filling also the steam-vessel, with a pressure approaching to that in the boiler, and equal to the weight of the pole and pump-rods. A comparatively small supply of steam from the boiler into the steam-vessel brought it up to the boiler pressure, sending the pole and pump-rods upwards with a spring. The steam-valve then closed, allowing the momentum of the great weight of pump-rods, together with the expanding steam, to complete the up-stroke. The discharge-valve was then opened for a moment, allowing a blast of steam to escape, reducing the pressure say to one-half. The weight of the rods caused their downward movement, raising the load of water in the plunger-pole pumps, and at the same time compressing the steam from the pole-case into the steam-vessel, equal at the finish of the stroke to the support of the pole and pump-rods. This most simple steam-engine combined in the greatest degree the two elements of expansion and momentum.

The up-stroke began with a much higher pressure of steam than was necessary to raise the load ; having given momentum to the rods, the supply of steam was cut off, and the stroke was completed by expansion. The down-stroke began with a comparatively low pressure of steam under the pole. The unsupported pump-rods fell downwards, setting in upward motion the column of water in the plunger pole pumps. The discharge-valve was closed long before the completion of the down-stroke, and the momentum of the moving mass of rods and water compressed the steam driven from the pole-case into the steam-vessel up to a pressure equal to the support of the pole and pump-rods. The pole was, therefore, continually floating or rising and falling in steam of ever-varying pressure.

Trevithick's figures show the working power of the 33-inch pole as much greater than Watt's 72-inch cylinder engine, even when the steam pressure in the former was much reduced, and that Woolf's double-cylinder engine, of less power, cost ten times as much as the pole-engine. This sum probably included the costly buildings required for the beam-engines, which Trevithick's plan dispensed with.

The reader may judge of the perfection of mechanism in this plain-looking engine from the fact that a pole, with 150 lbs. of steam to the inch in the boiler, was equal to 50 or 60 tons weight, thrown up and down its 10-feet stroke ten or fourteen times a minute, with a limit of movement perfectly under control, while modern engineers are building ships' turrets because of the difficulty of raising and depressing a 30-ton gun from the hold to above the water level.

[Rough draft.]

"Sir, "Camborne, *September 12th*, 1815.

"I received a letter dated the 20th of August, from Mr. Davies, in which he did not mention the name of Herland castings. On the 24th of the same month I wrote to you, informing you of the same, and requesting to know what state of forwardness the castings were in. On the 30th of August I received another letter from Mr. Davies, not saying what state of forwardness the castings were in, nor when they would be finished, only that they would set their hands about them, and that I might expect a letter from you stating the particulars, which has not yet come to hand. I have waited so long that I am quite out of patience. You will know that it is now nearly double the time that the castings were to have been finished in, and you have not yet answered my letters as to the state of the castings nor when they will be finished. I must again request you to write to me on this subject, otherwise I must immediately remove the orders to some other founders that may be a little more attentive to their customers. I must be informed in the positive, whether the castings will be at Bristol by the next spring-tide, as a vessel is engaged for the purpose of taking them to Cornwall.

"Yours, &c.,

"R. T.

"Mr. John Rastrick,
"Chepstow, South Wales."

Rastrick, whom he had known at the Thames Driftway, had become the managing engineer at the Bridgenorth Foundry

[Rough draft.]

"Mr. George Cowie, "Camborne, September 29th, 1815.

"I received your favour of the 20th, and on the 23rd called on Mr. Wm. Sims, your engineer, who went with me to Beeralstone Mine the same day. We arranged on the spot what was necessary for the engine. I hope it will be at work in good time, before the winter's floods set in. Nothing can prevent it, unless the castings are detained by contrary winds. The boilers are nearly finished in Cornwall. The castings at Bridgenorth are in a forward state. I intend leaving this evening for Bridgenorth, to ship the castings, both for Herland and Beeralstone. It was the wish of the agents on the mine that these castings might be sent to Swansea, and taken from thence to the mine with a freight of coal. I shall, if possible, get the Herland castings in the same ship. The workmen making your boilers want an advance of cash to enable them to finish. They provide both iron and labour, for which they are to receive 42l. per ton for the boiler when finished; the weight will be about 8 tons. You may send this money to Mr. Sims or to me, or otherwise you may direct it to Mr. N. Holman, boiler-maker, Pool, near Truro. 100l. will satisfy them for the present. I hope to be in London this day week, and will call at your office.

"Yours, &c.,

"R. Trevithick."

The pole-engine was not only used in several mines shortly after its first introduction, but Mr. Sims, the leading engineer of the eastern mines, not generally favourable to Trevithick, advocated its application in the traditional Watt district.

Scarcely had he smoothed the way with one opponent than another sprung up in an unexpected quarter. His brother-in-law, Harvey, with his once friend, Andrew Vivian, then a partner with Harvey, opposed his plans at the Herland. They were annoyed at Trevithick's sending his orders for castings and machinery to Bridge-north, and may have had doubts of the success of the new inventions. They had authority in the mine, pro-bably as shareholders, a position generally acquired in Cornwall by those who supply necessary mine material, as well as by the smelters who buy the mineral from the mines. The Williamses and Foxes, controlling the eastern district of mines, were also shareholders and managers, supplying machinery and buying the mine produce.

[Rough draft.]

" Sir,   " Penzance, 13th December, 1815.

        " Yesterday I was at Herland, where I was informed that Captain Andrew Vivian had been the day before, on his return from Mr. Harvey's, and discharged all the men on the mine, without giving them a moment's notice. Before the arrival of the castings the pitmen, sumpmen, carpenters, and smiths were very busy getting the pit-work ready ; at which time H. Harvey and A. Vivian were exulting in reporting that the iron ore was not yet raised that was to make the Herland castings. The day that they heard of their arrival they dis-charged all the labourers, and ordered the agents not to admit another sixpence-worth of materials on the adventurers' account, or employ any person whatever.

" The agents sent a short time since to Perran Foundry[1] for

---

[1] Belonging to Williams and Co.

the iron saddles and brasses belonging to the balance-bob, the property of the adventurers; but they refused to make them, with a great deal of ill-natured language about my engine.

"I am determined to fulfil my engagement with the adventurers, and yesterday ordered all the smiths, carpenters, pitmen, and sumpmen to prepare the adventurers' pit-work, and ordered the agents to get the balance-bob and every other thing that may be wanted at my expense, so as to fork the first lift, which I hope to have dry by Monday three weeks. The engine will be in the mine this week, and in one fortnight after I hope the engine will be at work, and in less than a week more the first twenty fathoms under the adit will be dry.

"In consequence of the Perran people refusing to send the saddles and brasses for the balance-bob, we will make shift in the best way we can without them. The brasses I have ordered on my own account at Mr. Scantlebury's. The coals for the smiths I have also ordered, and the same for the engine to fork the first lift. This is very uncivil treatment in return for inventing and bringing to the public, at my own risk and expense, what I believe the country could not exist without. I am determined to erect the engine at all events and upset this coalition before I leave Europe, if it detains me one year to accomplish it.

"I remain, Sir,

"Your very humble servant,

"RICHARD TREVITHICK.

"MR. PHILLIP,
"George Yard, Lombard Street.

"P.S.—I should be glad to hear from you what is going forward respecting an arrangement of the shares."

[Rough draft.]

"GENTLEMEN, "PENZANCE, 23rd December, 1815.

"I have received the Herland castings, and am very seriously sorry to say, after we had fixed together the castings on the mine and made the joints, on attempting to put the plunger-pole into the case it would not go down; neither would either of the rings go to their places into the cylinder and on to tae pole; therefore the whole engine must be again taken to pieces and sent to a turning and boring mill to be newly turned and bored. How to get this done I cannot tell, for the founders here will not do it because they had not the casting them. Already great expenses have been incurred by delays, and now to send them back to Bridgenorth at an immense loss of time and money will be a very serious business indeed. I think that either the cylinder is bored crooked or the plunger-pole turned crooked, or both, as it will sink farther down into the cylinder on turning it round on one side than it will on the other. The whole job is most shamefully fitted up, and was never tried together before sent off. Write to me by return of post and say what I am to do in this dilemma.

" Yours, &c.,

" RICHARD TREVITHICK.

" HAZELDINE, RASTRICK, AND Co.,
 " *Bridgenorth*."

The new engine-work from Bridgenorth on arrival was found to be so inaccurately made that the pole would not go into the pole-case. Henry Phillips,[1] who saw the engine make its first start, says :—

" I was a boy working in the mine, and several of us peeped in at the door to see what was doing. Captain Dick was in a great way, the engine would not start; after a bit Captain Dick threw himself down upon the floor of the engine-house, and there he lay upon his back; then up he jumped, and snatched a sledge-hammer out of the hands of a man who was driving in a wedge, and lashed it home in a minute. There never was a man could use a sledge like Captain Dick; he was as strong as a bull. Then, he picked up a spanner and unscrewed something, and off she went. Captain Vivian was near me, looking in at the door-way; Captain Dick saw him, and shaking his fist, said : 'If you come in here I'll throw you down the shaft.' I suppose Captain Vivian had something to do with making the boilers, and Captain Dick was angry because they leaked clouds of steam. You could hardly see, or hear anybody speak in the engine-

---

[1] Still working in Harvey's foundry at Hayle, 1869.

house, it was so full of steam and noise; we could hear the steam-puffer roaring at St. Erth, more than three miles off."

By the end of January, 1816, the engine was ready for work, and after ten days of experience, he thus described the result:—

"Mr. Davies Giddy, M.P.,   "Penzance, 11th February, 1816.

Sir,—I was unwilling to write you until I had made a little trial of the Herland engine. It has been at work about ten days, and works exceedingly well; everyone who has seen it is satisfied that it is the best engine ever erected. It goes more smoothly than any engine I ever saw, and is very easy and regular in its stroke. It's a 33-inch cylinder, 10½-feet stroke. We have driven it eighteen strokes per minute. In the middle, or about two-thirds of the stroke, it moved about 8 feet per second, with a matter in motion of 24 tons; and that weight returned thirty-six times in a minute, with 2 bushels of coal per hour. This of itself, without the friction, or load of water, is far more duty than ever was done before by an engine. I found that it required about 80 lbs. to the inch to work the engine the first twelve hours, going one-third expansive, twelve strokes per minute, 10½-feet stroke, with 24 bushels of coal. The load of water was about 30,000. This was occasioned by the extreme friction, the plunger-pole being turned, and the plunger-case bored, to fit so nicely from end to end, that it was with great difficulty we could at all force the plunger-pole down to the bottom of the plunger-case. This is now in a great degree removed, and since we went to work we have thrown into the balance-box 4 tons of balance, and it would carry 3 tons more at this time. We must have carried that load in friction against the engine, therefore, if you calculate this, you will find it did an immense duty, going twelve strokes per minute, 10½-feet stroke, with 2 bushels of coal per hour. The engine is now working regularly twelve strokes per minute, with 60 lbs. of steam, 10½-feet stroke, three-quarters of the stroke expansive, and ends with the steam rather under atmosphere strong, with considerably within 2 bushels of coal per hour. I would drive her faster, but as the lift is hanging in the capstan rope under water, they are not willing to risk it. I have raised the steam to 120 lbs. to the inch, the joints and every-thing perfectly tight. I took the packing out of the stuffing box and examined it, and found that the heat had not at all injured it; the packing is perfectly tight, not a particle of steam is lost.

"I have offered to deposit 1000*l*. to 500*l*. as a bet against Woolf's best engine, and give him twenty millions, but that party refuses to accept the challenge. I have no doubt but that by the time she is in fork she will do 100 millions, which is the general opinion here. The boilers are certainly the best ever invented, as well as the other parts. The draught is the best you ever saw; I have only one-quarter part of the fire-bars un-covered, yet from one-quarter part of the fire-place that I first made, I find plenty of steam. The greatest part of the waste steam is condensed in heating the water to fill the boiler; what escapes is a mere nothing. The engine will be loaded, when in fork, about 52 lbs. to the inch. Now suppose I raise the steam so high at the first part of the stroke as to go so expansive as to leave the steam, at the finish, only atmosphere strong, shall I, in that case, use any more coal than at present? The mate-rials and joints will stand far more than that pressure; 500 lbs. to the inch would not injure them. When the engine gets on two lifts, I will write to you again, and in the meantime please to give me your thoughts on the engine. Every engine that was erecting is stopped, and the whole county thinks of no other engine.

"Your very obedient servant,

"RD. TREVITHICK."

The new pole puffer-engine worked so satisfactorily and its movements were so manageable that the length of the stroke was increased by the spare 6 inches, which had been allowed as a margin in case of its overrunning its intended stroke. It would bear being worked at eighteen strokes a minute, while the Watt 72-inch engine did not exceed nine strokes a minute; with steam in the boiler of 80 lbs. to the inch it performed its work when the steam supply was cut off at two-thirds of the stroke, completing it by expansion. It also worked well with steam of 120 lbs. to the inch; but the want of strength in the pump-rods and the requirements of the mine caused the regular working pressure of steam to be reduced to 60 lbs. on the inch, and to be cut off when the pole had moved through the first quarter of its stroke. The excellent draught causing the fire-bars to be reduced to one-quarter of their original surface, and the heating the feed-water by the waste steam in this powerful pumping engine, indicate the use of the blast-pipe as at that time worked in the Welsh puddling-mill engine. Watt's engine was for a moment forgotten, that he might challenge Woolf to a trial, giving him as a help twenty millions, or the understood duty of the Watt engine. This non condensing pole-engine, with 20 tons of pump-rods, moved at a maximum speed of 8 feet a second, and was equal to its work with a steam pressure of 52 lbs. on the inch. Trevithick contemplated extending the expansive principle even further than he had done in the Wheal Prosper pole condensing engine, so that at the finish of the upstroke the steam should only be about the pressure of the atmosphere, or say from 1 to 10 lbs. on the inch, having commenced it with steam of from 100 to 200 lbs. on the inch, and cutting off the supply from the boiler when the pole had gone but a very small part of its upward stroke, more or less as the mine requirements admitted of it. The principle of expansive working and momentum of moving parts was of necessity modified in its application to pump-work.

" DEAR TREVITHICK,   " EASTBOURNE, *February* 15*th*, 1816.

   "I have been called here by the decease of my wife's uncle, and consequently your letters of the 11th did not reach me till this day.

   "The account you give me of your new engine has been

extremely gratifying. The duty performed by the engine in giving a velocity of 8 feet in a second, thirty-six different times in a minute, to 24 tons of matter, by the consumption of 2 bushels of coal in an hour, is indeed very great, amounting to about fifty-seven and three-quarter millions. So that when you obtain a proper burden, and the extraordinary friction arising from the too close fitting of the plunger-pole and case is reduced, there seems to be no doubt of your engine performing wonders.

"I am of opinion that the stronger steam is used, the more advantageous it will be found. To what degree it should be applied expansively must be determined by experience in different cases. It will depend on the rate at which the engine requires to be worked, and on the quantity of matter put into motion, so that as large a portion as possible of the inertia given in the beginning of the stroke may be taken out of it at the end.

"Some recent experiments made in France prove, as I am told, for I have not seen them, that very little heat is consumed in raising the temperature of steam. And if this is true, of course there must be a great saving of fuel by using steam of several atmospheres' strength, and working expansive through a large portion of the cylinder. I have really been impatient for a week past to receive some account of your machine, having learned nothing about it, except from a paragraph dated Hayle in the Truro paper of last Saturday week, and somehow or other the next paper has not reached me.

"I hope to be in London about Tuesday next, but at all events direct to me there, as my letters are regularly forwarded.

"Believe me, dear Sir,

"Yours ever most faithfully,

"DAVIES GIDDY."

[Rough draft.]

"MR. JOHN ADAMS,   "BROMSGROVE, 8th March, 1816.

"Sir,—I received your favour of the 12th February, but did not answer it in due course, because I was then erecting an engine on the new plan, which is now at work, and performs exceedingly well. It is equal in power to a 72-inch diameter cylinder, double power of B. and Watt's. The expense of erection, and the consumption of coals in this engine, are not one-third of a B. and Watt's to perform the same work. I am the same Trevithick that invented the high-pressure engine. I have sent out nine steam-engines to the gold and silver mines of Peru. I intend to sail for that place in about a month or six weeks, but shall appoint agents in England to erect these engines.

"No publication or description whatever has been in circulation, neither is it required, for I have a great many more orders than I can execute.

"I have not seen anything of Mr. Losh's patent engine, or Mr. Collins'.

"If you should go to London I advise you to call on Mr. Jas. Smith, Limekiln Lane, Greenwich, who is an agent for me, and will soon be able to show you an engine on this plan at work.

"I remain, &c.,

"R. TREVITHICK."

Unless the foregoing letters are based on error, the only conclusion to be drawn is that Watt, on the expiry of his patent right and of twenty-eight years of labour, having erected his masterpiece in Cornwall, was within a few years so beaten that Trevithick, in his challenge to Woolf, offered to throw in the Watt engine as a makeweight, and with such odds to bet him two to one that his comparatively small and cheap high-pressure engine should beat the two big ones, both in power, in first cost, and in economical working. The Watt engine was one of his largest, with a 72-inch cylinder. Its power was equal to Trevithick's 33-inch pole-engine, when worked with steam of 34 lbs. to the inch; but the latter also worked with three times that pressure of steam, whereby its power was increased threefold. The first cost of these engines was probably in inverse proportion to their power. Trevithick's cost 700*l.*, while

three times that sum would not pay for the Watt engine.
The reported duty of Watt's Herland engine was twenty-
seven millions; and if the trial was with his ordinary
bushel of 112 lbs. of coal, the duty would only be equal
to twenty millions with 84 lbs. of coal, which constituted the Cornish bushel.

Trevithick's pole high - pressure steam - engine did
fifty-seven millions; in other words, performed the same
work as the Watt engine with less than half of the daily
coal.   This large economy led to orders for many
engines, on his promise that they should cost much less
than those of Watt of equal power, and should perform
the work with one-third of the coal.   Some believed
him, though others were stony-hearted, and as obstinate
as donkeys.

[Rough draft.]

"Mr. Phillip,   "Penzance, 8th March, 1816.

   "Sir,—I long since expected to have heard from you that
my agreement with the Herland adventurers was executed. I
have in every respect fulfilled my part of the engagement with
the adventurers, and expect that they will do the same with
me. The engine continues to work well. Every person that
has seen it, except Joseph Price, A. Vivian, Woolf, and a few
other such like beasts, agrees that it is by far the best engine
ever erected. Its performance tells its effects, in spite of all
false reports.

   "Joseph Price and A. Vivian reported that the engine was
good for nothing, that it would not do four millions, and that
at the next Tuesday meeting they would turn it idle. On the
evening before the meeting they met at Camborne for that
purpose.

   "Captain A. Vivian did not attend the meeting. I could
not help at the meeting threatening to horsewhip J. Price for
the falsehoods that he with the others had reported.

   "I hear that he is to go to London to meet the London
committee on Monday. I hope the committee will consider

J. Price's report as from a disappointed man. It is reported that he has bought very largely in Woolf's patent, which now is not worth a farthing, besides losing the making my castings, which galls him very sorely.

" The water sinks regularly 20 fathoms per month, including every stoppage. On Monday next I expect they will be putting down the second lift. The water rises about 8 inches per hour when the engine is idle, and when at work will sink it again at the same rate, showing that the engine is equal to double the growing stream. When drawing from the pool the sinking is not much above 4 inches per hour, which shows that the water drains from a great distance from the country. The engine is going fourteen strokes per minute, 10-feet stroke, 14½-inch box. When Herland worked last they drew a 14-inch box, 7-feet stroke, twelve strokes per minute in winter, and seven strokes per minute in summer. Therefore it appears that the winter water is about from seven to eight strokes per minute, and the summer water from four to five strokes per minute for this engine.

" The engine has forked faster the last week than she did before. I think that the great quantity of water that was laying round the mine at the surface is nearly drawn down, and that as we get down to a closer ground the drainage will not be so much. If we have dry weather the water will, at the next shallow level, fall off two strokes per minute before the next lift is in fork. If it continues the same we can continue to sink 20 fathoms per month, exclusive of the time it will take to fix the lifts. As we get down the house of water will lessen considerably. The expense of the engine is about 100*l.* per month. The sumpmen and others attending on the forking the water, about 100*l.* per month more. They have all the materials on the mine for the pit-work, therefore a very trifling sum will bring the water down to the 60-fathom level, when the mine will pay her own expense.

" I will thank you for an account of the meeting.

" Your obedient servant,

" R. TREVITHICK."

Mr. Phillip was the financial managing shareholder —more particularly with the Londoners—at that resuscitation of Herland Mine; and though the new engine was comparatively cheap, both in its first cost and in its consumption of coal, and satisfactorily reduced the water in the mine, payment for it was withheld because the currents of self-interest were against Trevithick. Mr. Joseph Price was the manager of a steam-engine manufactory at Neath Abbey, in South Wales, and had been in the habit of supplying castings for Cornish mines. Arthur Woolf was then striving to bring into use his patent double-cylinder engine, and patent high-pressure steam-boiler, which Trevithick looked on as copies from Hornblower and himself. This, added to Woolf's sarcastic manner of speech, roused Trevithick's anger.

Putting aside the words of the disputants, the fact is stated that the pole-engine, with a reduced steam pressure, worked a pump $14\frac{1}{2}$ inches in diameter, 10-feet stroke, fourteen strokes per minute; while the largest and best engine by Watt in Cornwall, placed on the same mine, with a 72-inch steam-cylinder, gave motion to a pump of 14 inches in diameter, 7-feet stroke, at twelve strokes a minute; being in round numbers just one-half the amount of the work performed by Trevithick's comparatively small engine, which had not a single feature of the Watt engine in it.

[Rough draft.]

" CAPTAIN JOE ODGERS,   " PENZANCE, *March* 7,.1816.

"Sir,—I have your favour of the 27th February, and requested Mr. Page to send to you a sketch of the agreement. On seeing him yesterday, I found that he had neglected to send it to you. He will leave the country for London in a few days,

and intends to call on you at Dolley's as soon as he arrives. I
do not know that the agreement matters much for a few days
up or down. The terms are well understood between us, which
is that the adventurers and I equally share the advantages
that may arise from this new engine over Boulton and Watt's.
When you have fully arranged with your adventurers about the
engine, please to write me, and I will immediately proceed
to order the engine; and in the interim the agreement will be
drawn up by Mr. Page, and executed either here or in London,
just as may suit.

" I remain, &c.,

" R. TREVITHICK.

"P.S.—Herland engine goes on better and better. Your
adventurers will get a satisfactory account by applying in town
to Mr. Wm. Phillip, No. 2, George Yard, Lombard Street. He
is the principal of the London adventurers."

Trevithick believed that mine adventurers would
agree to pay him one-half the saving caused by his
engines, as compared with the cost of fuel in the Watt
engine. The duty performed by the latter was understood and agreed to generally; persons were chosen by
the adventurers to experiment and report on the duty of
Trevithick's pole-engine, that the amount of payment
might be ascertained in proportion to the saving effected.

" MR. GIDDY,   " PENZANCE, *April* 2, 1816.

" Sir,—I have long wished to write to you about the
Herland engine, but first wished to see the engine loaded with
a second lift, and a trial made of the duty. Yesterday was
fixed on, before ordering another engine for the eastern shaft.

" The persons attended. The arbitrators gave the duty as
forty-eight millions, and said they had no doubt the engine
would perform above sixty millions before getting to the bottom
of the mine.

" They were much within the duty, but I did not contend
with them, as they said it was quite duty enough.

H 2

"The engine worked 9¼ strokes per minute, with 2 bushels of coal per hour for the whole time, 10-feet stroke. There were two pump-lifts of 14½-inch bucket, making 43 fathoms, and 26 fathoms of 6-inch for house-water.

"The steam was from 100 to 120 lbs. to the inch. The valve open while the plunger-pole ascended 20 inches, then went the remainder of the 10-feet stroke expansive.

"It went exceedingly smooth and regular. Some time since, by way of trying the power of the engine, we disengaged the balance-bob. The engine worked twenty strokes per minute, with 17 tons of rods, &c., and drew 14½-inch bucket 23½ fathoms, and a 6-inch bucket 26 fathoms, 10-feet stroke, twenty strokes per minute.

"This was about 45,000 lbs. weight, with the speed of 200 feet per minute, which makes the duty performed more than the power of three 72-inch cylinders, single, of Boulton and Watt, say of 8-feet stroke, 10 lbs. to the inch, nine strokes per minute, which is more than these engines will perform. I have all the orders for every engine now required in the country, which is not to be wondered at, for one-tenth part of the expense in the erection will do, and the duty is not less than three times as much as other engines. This will be proved before we get to the bottom.

"The engine now works at about two-fifths of the load which she will have when at the bottom. When the next lift is in fork I will write to you again.

"I am, Sir, your humble servant,

"RICHARD TREVITHICK."

Independent examiners reported that the Herland pole-engine did forty-eight millions of duty, under various pressures of steam, up to 120 lbs. to the inch, working five-sixths of the stroke expansively, with a speed of twenty strokes a minute, or double the speed of the Watt engine; and the importance of those facts deserves the scrutiny and close study of youthful engineers. A small cheap engine, of 33-inch cylinder, similar in general construction to the Wheal Prosper pole-engine, but still more simple from the absence of air-pump and condenser, did as much work as three of Watt's largest engines with cylinders of 72 inches in diameter. This great stride in the useful value of the steam-engine was forced on the public by Trevithick's single-handed energy, when every man was against him, even Henry Harvey, his brother-in-law and friend, his former partner, Andrew Vivian, and his once carpenter and assistant, Arthur Woolf. As a closing attempt to finally crush him, he was made personally responsible for the payment of an engine erected for the benefit of others. This was the great trial test of the power and economy of the purely high-pressure expansive steam-engine as compared with the Watt low-pressure vacuum engine applied to large pumps.

" SIR,   " HAYLE FOUNDRY, 18th *April*, 1816.

"I was at Herland to-day. Captain Grose received a letter while I was there, signed by Captain William Davey and Joseph Vivian, requesting him to appoint others to attend the trial of the engine, as it would not be convenient for them to do it.

"Captain Samuel Grose, jun., brought down the drawing for your engine. He said he had taken off the working gear only. If you would wish it, we will make the working-gear and all the wrought-iron work on the drawing for the two engines ordered, and will take on a man or two immediately for that purpose. You will let me know about this before you set off, and also if any alteration is to be made in the beam for Wheal Treasure engine, since you have altered the size of the pole. We had cast the case, but I suppose it will suit some place else.

" Your obedient servant,

" HENRY HARVEY.

" MR. RICHARD TREVITHICK."

"About 1815 or 1816 I was employed by Captain Trevithick to erect various pole-engines, one of them at Saltram Stream. It had worked at Tavistock; it was a horizontal high-pressure pole puffer. Captain Samuel Grose was then erecting for Captain Trevithick a 24-inch high-pressure pole-engine at Beeralstone, on the Tamar, to drain a lead mine. I assisted Captain Grose. The stroke was about 8 feet. It worked with cross-head and side rods. There were two wrought-iron boilers about 3 feet 6 inches in diameter and 40 feet long. The fire and flues were outside. The steam pressure, 60 lbs. to the inch. I also erected a similar engine with a 20-inch pole at Wheal Treasure, now called Fowey Consols Mine; and one at Legassack, near Padstow. Those two had brass poles. It was found that the poles cut and wore in their passage through the stuffing box, the middle wearing more than the ends, causing steam to escape. A similar pole of Captain Trevithick's erecting was then working at Wheal Regent, near St. Austell.

"In 1818 I saw working at Wheal Chance Mine, near Scorrier, an old 60-inch cylinder Boulton and Watt engine. A pole of Trevithick's was fixed between the cylinder and the centre of the main beam. High-pressure steam was first worked under the pole and then expanded in the cylinder."[1]

The late Mr. William Burral, for many years manager of the boiler-making department at Messrs. Harvey and Co., at Hayle, said:—

"About the year 1815 or 1816 I helped to erect at Treskerby Mine an engine for Captain Trevithick. Mr. Sims was the engineer of the mine. The engine had the usual cylinder, and close to it one of Captain Trevithick's poles was fixed. The boilers were Captain Trevithick's high-pressure. The steam was first turned on under the pole. When she had finished her up-stroke the steam passed from under the pole on to the top of the piston in the cylinder. There was a vacuum under the piston. The steam-cylinder was 58 inches in diameter, about 9 or 10 feet stroke. The pole was 36 inches in diameter, and a less

---

[1] Recollections of Captain H. A. Artha, Penzance, 1870.

stroke than the piston, because it was fixed inside the cylinder, nearer to the centre of the beam. There was a pole-engine then working at Wheal Lushington, also at Poldice, and at Wheal Damsel."

"Captain Artha recollects at Wheal Alfred Mine in 1812 the 66-inch cylinder pumping engine used a pole air-pump; one or two whim-engines on the same mine also used them. Wheal Concord pumping engine, in 1827 had a similar air-pump. Old Wheal Damsel, near Treloweth, used one as late as 1865. The condensing water and air passed through a branch with a valve on it near the top of the pole-case, just under the stuffing box There was a foot-valve at the bottom of the pole-case."[1]

The writer has had the pleasure of personal acquaintance with each of those three gentlemen, who as young engineers commenced their labours in the erection of Trevithick's engines.

No sooner had Trevithick perfected the pole condensing engine and then the pole puffer-engine, than he, in conjunction with Sims, who had just taken part in the erection of one of his high-pressure steam pole-engines for working the pumps at Beeralstone Mine, combined the pole with the ordinary Watt vacuum engines, supplying them with steam from his high-pressure boilers, in other words, converting them from their original form of low-pressure vacuum engines to high-pressure expansive compound steam-engines.

The old 60-inch cylinder Boulton and Watt engine, at Wheal Chance (one of Watt's favourite engines), was in 1818 transformed into a high-pressure engine, with Trevithick's pole placed between the centre of the main beam and the steam-cylinder. The high-pressure steam from Trevithick's new boilers was turned under the

---

[1] Captain Artha became the resident engineer at the Real del Monte mines in Mexico; Captain Samuel Grose, one of the first Cornish mine engineers; and Mr. Burral, the engineer of a department at the engineworks of Messrs. Harvey and Co.

pole for the up-stroke, after which it was expanded in
the old and much larger cylinder on the top of the
piston causing the down-stroke; it then, by its passage
through the equilibrium valve, allowed the piston in the
large cylinder to make its up-stroke, by equalizing
the pressure of steam on its top and bottom, while a
fresh supply of strong steam from the boiler admitted
under the pole gave power to the up-stroke; and finally,
the comparatively low-pressure steam under the large
piston passed to the condenser and air-pump to form a
vacuum for the down-stroke, as in the Watt engine.

Sims, the engineer at Wheal Chance, one of the mines
in the eastern or Watt district, was converted and became
in 1815 or 1816 a partner with Trevithick, and erected,
at Treskerby Mine, Trevithick's high-pressure pole of
36 inches in diameter, as an addition to the old Watt
engine working with a cylinder 58 inches in diameter.

Watt, then, within a year or two of his death, was too
old to any longer take part in the contest; his engine
in the hands of others was converted and became a
high-pressure expansive engine.

Trevithick, as a further proof that he could do without the Watt patent air-pump bucket, with its piston
and valves, removed it from a Watt engine at Wheal
Alfred Mine in 1812, replacing it by one of his poles,
answering the same purpose, but different in construction. Many other mines used them; one remained at
work in Old Wheal Damsel in 1860. They have also
been used in steamboat air-pumps.

Having traced during a period of five or six years
the rise and progress of the high-pressure expansive
pole condensing-engines, the high-pressure expansive
pole puffer-engine, and the combined pole and cylinder
high-pressure engine, their value in a commercial sense

may be further tested by the public acts of the time. Lean, an authority on such matters, and certainly not given to unduly praise Trevithick, spoke as follows on the duty of those particular engines at various periods; and not the least noteworthy is the fact, that Herland, Poldice, and Treskerby, that were prominent in the early use of the Watt engine, threw off their allegiance but shortly before the last days of the great engineer, and converted his low-pressure steam vacuum engines into Trevithick high-pressures.

"In 1798 Messrs. Boulton and Watt, who on a visit to Cornwall, came to see it—'the Herland engine'—and had many experiments tried to ascertain its duty; it was under the care of Mr. Murdoch, their agent in the county. Captain John Davey, the manager of the mine, used to state that it usually did twenty millions, and that Mr. Watt, at the time he inspected it, pronounced it perfect, and that further improvement could not be expected.

"In 1811 the average duty of the three engines (Boulton and Watt's) on Wheal Alfred Mine was about twenty millions. These engines were at that time reckoned the best in the county.

"In 1816 Sims erected an engine at Wheal Chance, to which he applied the pole adopted by Trevithick in his high-pressure engine. This engine attained to forty-five millions; and in 1817 it did 46 9 millions.

"In 1814 Treskerby engine is reported as doing 17·48 millions.—Wm. Sims, engineer.

"In 1820 Treskerby engine, to which Trevithick's high-pressure pole had been adapted, had reached 40·3 millions."[1]

The Herland engine of Watt in 1798 did twenty millions; in 1816 Trevithick's high-pressure pole puffer in the same mine did forty-eight millions. In 1820 his high-pressure pole-engine was combined with a

[1] Lean's 'Historical Account of the Steam-Engine in Cornwall,' pp. 11, 32, 36.

Watt low-pressure engine, thereby more than doubling its economical duty.

In 1813 Trevithick wrote :—

"That new engine you saw near the sea-side with me (Wheal Prosper high-pressure pole condensing engine) is now lifting forty millions one foot high, with a bushel of coal, which is nearly double the duty that is done by any other engine in the county. A few days since I altered a 60-inch cylinder engine at Wheal Alfred to the same plan, and I think she will do equally as much duty. I have a notice to attend a mine meeting, to erect a new engine equal in power to a 63-inch cylinder single."[1]

In the four or five years from his return to Cornwall in 1810, to his leaving for South America in 1816, he doubled the duty and the power of the steam-engine. Watt once said he had received an oblique look from Trevithick, sen. The time was now come for Trevithick, jun., to return the compliment; his improved engines having made their way into the eastern mine district, which Watt once looked upon as his own.

Trevithick was short of money and on the point of leaving England for South America, when Mr. Sims, in the employ of Messrs. Williams and Co., favourable to low pressure, was sent to negotiate for the purchase of a share in Trevithick's patent of 1815 for the high-pressure steam expansive pole-engine.

"18th October, 1816.—Agreement between Trevithick and Mr. William Sims, prepared by myself and Mr. Day, solicitor for Mr. Sims, or for Mr. Michael Williams, under whom Sims acted, recites, that in consideration of 200l. paid by Sims, he was to have a moiety of the patent for Cornwall and Devon, and that I should have power to act and make contracts whilst Trevithick was out of England.

"The day after contract signed, Trevithick sailed in the

---

[1] See Trevithick's letter, January, 1813, vol. ii., p. 55.

'Asp,' Captain Kenny, for South America. I was on board when the ship sailed.

"I see among my papers, in May, 1819, in reference to the patent, is the following note:—'Mr. Michael Williams said it was verbally agreed that Captain Trevithick should have one-quarter part of the savings above twenty-six millions.' This, I believe, was the average duty of the engines at that time.

"I had several assurances relative to Trevithick's claims, and much correspondence, but no allowance was made from any mines but Treskerby and Wheal Chance; though Trevithick's patent and boilers were used throughout the county without acknowledgment; and the duty of the engines had soon increased from twenty-six millions to about seventy millions.

"In 1819 I attended at the account-houses of Treskerby and Wheal Chance, of which the late Mr. John Williams, of Scorrier, was the manager, in consequence of some of the adventurers objecting to continue the allowances on the savings to Captain Trevithick, when Mr. Williams warmly observed, that whatever other mines might do, he would insist, as long as he was manager for Treskerby and Wheal Chance, the agreement made should be carried into effect.

"I remain, my dear Thomas,

" Your very affectionate father,

" RD. EDMONDS."[1]

The agreement with Mr. Sims, or rather with Mr. Michael Williams, late M.P. for Cornwall, who exercised large authority in Cornish mines, was that he should have for 200*l.* one-half of the patent for the high-pressure pole-engine, as applied to Cornwall and Devon.

Trevithick had desisted from securing a patent for the large high-pressure steam-boilers and expansive working, on a verbal understanding that he should receive one quarter of the saving from the reduced con-

---

[1] Portion of a letter written at Penzance, 8th February, 1853.

sumption of coal by those two particular inventions, twenty-six millions of pounds of water raised one foot high by a bushel of coal of 84 lbs., being the duty of the best Watt engines, to be taken as a starting-point for the payment. Treskerby and Wheal Chance paid for the pole-engine, but the Trevithick boilers suitable for high steam, and the simple methods of working it expansively, had been made so generally public, that people professed to think they had a right to them, when but a few years before they had thrown the inventor off his guard by saying "everybody knows that the Cornish boiler is your plan, and as it cannot be denied, a patent will be of no service."

Mr. John Williams[1] stated "that whatever other mines might do, he would insist, as long as he was manager for Treskerby and Wheal Chance, the agreement made should be carried into effect." The Williamses paid to Trevithick 300*l*. for the saving of coal by the pole patent engine, as an "acknowledgment of the benefits received by us in our mines;" but no payment was made for the greater invention of the high-pressure steam-boilers then in general use.

In 1814 the Watt Treskerby engine did seventeen and a half millions. Trevithick's boiler and pole were applied, and the duty was increased to more than forty millions. In 1816 the same changes were made in Wheal Chance, and the duty rose to more than forty-six millions. The consumption of coal was reduced to one-half, amounting in round numbers to a gain of 500*l*. a month in those two mines alone.

---

[1] Mr. John Williams had the remarkable dream, many hours before the event, enabling him to describe the particulars of the assassination of Perceval in 1812.

"TREVINCE, near TRURO,

" DEAR SIR,   5th January, 1853.

" I am favoured with your letter of the 31st ult., enclosing
also one from Mr. F. Trevithick, of the 24th idem, and have
much pleasure in complying with your joint request to the best
of my ability. I was well acquainted with the late Mr. Rd.
Trevithick, having had frequent occasion to meet him on business and to consult him professionally; and I am gratified in
having the present opportunity of bearing testimony to his distinguished abilities, and to the high estimation in which the
first Cornish engineers of the day then regarded him. I need
scarcely say that time has not lessened the desire in this county
especially to do him justice. As a man of inventive mechanical
genius, few, if any, have surpassed him, and Cornwall may well
be proud of so illustrious a son.

" At this distance of time I can scarcely speak with sufficient
exactness for your purpose of the numerous ingenious and valuable mechanical contrivances for which we are indebted to him,
but in reference to his great improvements in the steam-engine
I have a more particular recollection, and can confidently affirm
that he was the first to introduce the high-pressure principle of
working, thus establishing a way to the present high state
of efficiency of the steam-engine, and forming a new era in the
history of steam-power. To the use of high-pressure steam, in
conjunction with the cylindrical boiler, also invented by Mr.
Trevithick, I have no hesitation in saying that the greatly-
increased duty of our Cornish pumping engines, since the time
of Watt, is mainly owing; and when it is recollected that the
working power now attained amounts to double or treble that
of the old Boulton and Watt engine, it will be at once seen that
it is impossible to over-estimate the benefit conferred, either
directly or indirectly, by the late Mr. Trevithick, on the mines
of this county. The cylindrical boiler above referred to effected
a saving of at least one-third in the quantity of coal previously
required; and in the year 1812 I remember our house at Scorrier paying Mr. Trevithick the sum of 300l. as an acknowledgment of the benefits received by us in our mines from this
source alone. Mr. Trevithick's subsequent absence from the

county, and perhaps a certain degree of laxity on his own part in the legal establishment and prosecution of his claims, deprived him of much of the pecuniary advantage to which his labours and inventions justly entitled him; and I have often expressed my opinion that he was at the same time the greatest and the worst-used man in the county.

"Amongst the minor improvements introduced by him, it occurs to me to notice that he was the first to apply an outer casing to the cylinder, and by this means prevent, still further than Watt had succeeded in doing, the loss of heat by radiation.

" As connected with one of the most interesting of my recollections of Mr. Trevithick, I must mention that I was present by invitation at the first trial of his locomotive engine, intended to run upon common roads, and of course equally applicable to train and railways. This was, I think, about the year 1803, and the locomotive then exhibited was the very first worked by steam-power ever constructed.

"The great merit of establishing the practicability of so important an application of steam, and the superiority of the high-pressure engine for this purpose, will perhaps more than any other circumstance serve to do honour through all times to the name of Trevithick. The experiment which was made on the public road close by Camborne was perfectly successful; and although many improvements in the details of such description of engines have been since effected, the leading principles of construction and arrangements are continued, I believe, with little alteration, in the magnificent railroad-engines of the present day. Of his stamping engine for breaking down the black rock in the Thames, his river-clearing or dredging machine, and his extensive draining operations in Holland, I can only speak in general terms, that they were eminently successful, and displayed, it was considered, the highest constructive and engineering skill. As a man of enlarged views and great inventive genius, abounding in practical ideas of the greatest utility, and communicating them freely to others, he could not fail of imparting a valuable impulse to the age in which he lived; and it would be scarcely doing him justice to limit his claims as a public benefactor to the inventions now clearly traceable to him, important and numerous as these are. From my own impressions I may say that no one could be in his presence without being struck with the originality and richness of his mind, and without deriving benefit from his suggestive conversation. His exploits and adventures in South America, in connection with the Earl of Dundonald, then Lord Cochrane, will form an interesting episode in his career; and altogether, I am of opinion that the Biography which you have undertaken will prove highly interesting and valuable, and I wish you every success in carrying it out.

"Believe me, my dear Sir,

"Yours very faithfully,

"MICHAEL WILLIAMS.

"E. WATKIN, Esq.,
" *London and North-Western Railway,*
" *Euston Station, London.*"

Arthur Woolf shortly after that time (1811) erected his double-cylinder engines in Cornwall. The late Captain Samuel Grose, when giving the writer his recollections of Trevithick, said :—

"When he returned from London to Cornwall, about 1810 or 1811, he employed me to look after the erection of the Wheal Prosper high-pressure engine. Oats, Captain Trevithick's head boiler-maker, was constructing the boilers; Woolf came into the yard, and examined them. 'What do'st thee want here?' asked Oats. 'D—n thee, I'll soon make boilers that shall turn thee out of a job!' was Woolf's reply. He was a roughish man. When his brother Henry mutinied at the Nore, Woolf, who was then working an engine in Meux's brewery, and had married the lady's maid, made interest with his employer to save Henry from being hanged at the yard-arm, and afterwards found employ for him in Cornwall. He was but a clumsy mechanic. Woolf used to blow him up by saying, 'D—n thee, I wish I'd left thee to be hanged.'"

The writer, who knew Oats, has heard him tell similar stories of the rival engineers.

In 1800, Woolf, who had been a mine carpenter, went to London with the first high-pressure steam-engine which Trevithick had sent beyond the limits of Cornwall[1]—probably to Meux's brewery,[2] for he was there in 1803, and in the receipt of 30*l.* a year from Trevithick as engine-fireman. From the date of Woolf's patent in 1804, his pay from Trevithick ceased, and with it their friendship. Trevithick used to say, "Woolf is a shabby fellow."

Patents sprang up like mushrooms after Trevithick had so liberally cast forth the seeds of the high-pressure engine, making the security, or even the form of a patent, a doubtful matter. The perfecting of expansive high-pressure engines was like the boiler, the result of years of trial. When matured in 1816 it saved Cornwall and the world one-half of the coal that before had been consumed in low-pressure steam-engines. Every engineer became, more or less, an expansive worker, and Trevithick's saving of hundreds of thousands of pounds annually to the general public, gave to him little or no reward.

At the period of those high-pressure pole-engine experiments, Trevithick had devoted twenty years of constant labour to the improvement and extended use of the steam-engine, causing it to assume every variety of form except that of the Watt patent engine, an approach to which was unusual, as evidenced in the high-pressure steam Kensington model of 1796, without beam, parallel motion, air-pump, or condenser, having no one portion either in principle or detail similar to the Watt engine, being portable and not requiring condensing water, with single and double cylinders,

---

[1] See Trevithick's account-book, vol. i., p. 90.
[2] Captain John Vivian's recollections, vol. i., p. 142.

placed vertically or horizontally. Having during twelve busy years constructed over a hundred high-pressure steam-engines, scarcely any two of which were exactly alike, he departed if possible still further from the Watt type, and went back apparently, though not in reality, to the Newcomen engine, simplifying it by the omission of the great bob, and use of condensing water, as in the nautical labourer and steamboat engine of about 1810,[1] and the South American mine engines of 1816,[2] which had open-top cylinders, more like a Newcomen than a Watt, but if possible even more simple and primitive-looking than the former. Again, compare the thrashing engine of 1812[3] with the Newcomen of 1712:[4] the great and all-important difference being that one was a high-pressure steam-engine, the other a low-pressure atmospheric engine. Then came the varieties of high-pressure steam pole-engines, working very expansively either as puffers or condensers, retaining the same dissimilarity to the Watt engine: and lastly, the combination of the high-pressure pole with the Watt patent engine, thereby causing the old Watt engine to do more than double the work it had done when new from the hands of the maker, and also to perform this increase of work with a decrease in the consumption of coal.

The following chapter will trace the adaptation of high-pressure expansive steam, from cylindrical boilers, to the form of pumping engine still in general use.

---

[1] See vol. i., p. 356.   [2] See chap. xxi.   [3] Vol. ii., p. 37.
[4] Vol. i., p. 5.

CHAPTER XX.

THE WATT AND THE TREVITHICK ENGINES AT DOLCOATH.

HAVING up to 1816 traced the progress of the steam-engine in Cornwall through a century, during the latter half of which Trevithick, sen., and his son were among its most prominent improvers, the latter having devoted a quarter of a century to the work, the effect of which is shown in the skeleton outlines of a few classes of engines, one important feature still remains for examination before a correct judgment can be formed of the events of this period and their prime movers.

The use of an increasing pressure of steam gave increased force and value to the improved steam-engine, but the power of constructing engines and boilers to render the increased pressure manageable was the result of a lifetime of labour.

Savery, whose engine was scarcely more than a steam-boiler, failed to control its force, and is said to have blown the roof from over his head. The mechanism of Newcomen's engine was well arranged, but suitable only for the working of pumps, and its power was limited to the weight of the atmosphere, from which it was called the atmospheric engine.

In 1756, an atmospheric engine with a cylinder of 70 inches in diameter worked at the Herland Mine, " the only objection to which was the cost of the coal, to lessen which several methods had been suggested for increasing the elasticity of the steam, and reducing the size of the boiler."[1]

In 1775 Richard Trevithick, sen., removed the flat top of a Newcomen boiler, and substituted a semi-circular top, enabling it to contain stronger steam, and at the same time he improved the mechanical part of the engine by finding a better resting-place for the steam-cylinder than the top of the large boiler. Pryce gives a drawing of this engine as the best at that time in Cornwall.[2]

"It is known as a fact that every engine of magnitude consumes 3000l. worth of coal every year.

"The fire-place has been diminished and enlarged again. The flame has been carried round from the bottom of the boiler in a spiral direction, and conveyed through the body of the water in a tube (one, two, or three) before its arrival at the chimney.

"Some have used a double boiler, so that fire might act on every possible point of contact, and some have built a moorstone boiler, heated by three tubes of flame passing through it.

"A judicious engineer does not attempt to load his engine with a column of water heavier than 7 lbs. on each square inch of the piston."[3]

While Pryce's book was being printed, Watt in 1777 wrote of the Cornish steam-engines :—

"I have seen five of Bonze's engines, but was far from seeing the wonders promised. They were 60, 63, and 70 inch cylinders at Dolcoath and Wheal Chance. They are said to use each about 130 bushels of coals in the twenty-four hours, and to make about six or seven strokes per minute, the stroke being under 6 feetea ch. They are burdened to 6, 6½, and 7 lbs. per inch."[4]

---

[1] Borlase's ' Natural History of Cornwall.'
[2] See drawing, vol. i., p. 25.
[3] See Pryce's ' Mineralogia Cornubiensis,' published 1778.   Appendix.
[4] Smiles' ' Lives of Boulton and Watt.'

The 63-inch was an open-top cylinder atmospheric engine at Dolcoath Mine under the management of Trevithick, sen.; and shortly after, in 1777 or 1778, Watt's first engine was erected in Cornwall.[1]

In 1783 Trevithick, sen., gave Watt an order for a patent engine for Dolcoath, in size similar to the old Newcomen atmospheric, having a cylinder 63 inches in diameter, that a working trial might be made between the rival engines. The Watt engine having a cylinder-cover, with the patent air-pump and condenser, was known in the county as the Dolcoath great 63-inch double-acting engine. Three steam-engines were then at work in that mine: Trevithick senior's Carloose (then called Bullan Garden) atmospheric 45-inch cylinder, the atmospheric 63-inch cylinder, and Watt's 63-inch cylinder double-acting vacuum engine; all of which continued in operation side by side for five years until 1788, when for a time Dolcoath ceased to be an active mine. Trevithick, jun., was then a boy of seventeen years.

After ten years of idleness and rust, as if mourning the death of Trevithick, sen., in 1798 Richard Trevithick, jun., as engineer, and Andrew Vivian as manager, induced shareholders to resuscitate the old mine. Fire was again given to the voracious jaws of the boilers, and the three engines recommenced their labours and their rivalries.

A year or two before this Trevithick had made models of high-pressure steam-engines. Davies Gilbert, in 1796, met him among other engineers, giving evidence in the Watt lawsuits, when he mentioned his ideas of an engine to be worked solely by the force of steam. Watt had claimed such an engine in his patent

---

[1] See vol. i., p. 30.

twenty-seven years before, but had failed to carry it into practice. Hornblower had tried something like it in his double-cylinder expansion engine, but he did not use high-pressure steam, and consequently also failed.

The *idea*, therefore, of expansive steam was not new, but the *useful mastery* of it was. Savery had tried expansive steam before Watt patented it; the latter went to law with Hornblower for an infringement of the *idea*, when neither of them had in truth constructed an expansive steam-engine. The low pressure of the steam from the boilers used by Hornblower and Watt did not admit of profitable expansion in the cylinder; at its full boiler pressure it constituted but a comparatively small portion of the power of the engine : to reduce that power by expansion was as apt to be a loss as a gain. The steam-engine was still dependent for its power mainly on steam as an agent for causing the required vacuum, until 1796, when Trevithick disclosed his method of constructing small cylindrical boilers and engines suitable for giving power from the strong pressure of the steam, irrespective of vacuum.

Lean, who favoured Watt rather than Trevithick, thus records the advent of Watt's expansive engine :—

"In 1779 to 1788 Mr. Watt introduced the improvement of working steam expansively, and he calculated that engines which would previously do nineteen to twenty millions would thus perform twenty-six millions; but I do not find any record of this duty being performed in practice. In 1785 Boulton and Watt had engines in Cornwall working expansively, as at Wheal Gons and Wheal Chance in Camborne; but in these the steam was not raised higher than before, and the piston made a considerable part of the stroke therefore before the steam-valve was closed.

"In 1798, on account of a suit respecting their patent, which was carrying on by Boulton and Watt, an account of the duty of all the engines in Cornwall was taken by Davies Gilbert, Esq., and the late Captain Jenkin, of Treworgie, and they found the average to be about seventeen millions."[1]

One of these so-called expansive Watt engines, erected at Wheal Chance, was converted into a real expansive engine by Trevithick, as described in the foregoing chapter, by his high-pressure steam-boilers and the addition of his pole-engine. The conversion of the other, a 63-inch low-pressure vacuum engine at Wheal Gons, will be traced in this chapter.

Mr. Taylor, who for many years took an active interest in Cornish mining, says :—

"In 1798 an engine at Herland was found to be the best in the county, and was doing twenty-seven millions, but being so much above all others, some error was apprehended. This engine was probably the best then ever erected, and attracted therefore the particular attention of Messrs. Boulton and Watt, who, on a visit to Cornwall, came to see it, and had many experiments tried to ascertain its duty. It was under the care of Mr. Murdoch, their agent in the county.

"Captain John Davey, the manager of the mine, used to state that it usually did twenty millions, and that Mr. Watt, at the time he inspected it, pronounced it perfect, and that further improvement could not be expected."[2]

This best engine from the hands of Watt and Murdoch in the Herland Mine in 1798 may be taken as a Watt stand-point, when its usual duty was twenty millions; and Trevithick and Bull erected a competing engine, probably with an increased steam pressure, for Trevithick's portable high-pressure engines were at that time coming into notice;[3] but no trace remains of the result of this contest of the Watt and the Bull engine, though it was one of the causes of the lawsuits.

---

[1] Lean's 'Historical Statement of Steam-Engines in Cornwall,' p. 7.
[2] 'Records of Mining,' by John

Taylor, F.R.S., &c., part i., p. 155; published 1829.
[3] See vol. i., p. 95.

" In 1799 Henry Clark worked as a rivet boy in Dolcoath, and carried rivets to construct Captain Trevithick's new boiler, said to be the first of the kind ever made. It looked like a great globe about 20 feet in diameter, the bottom hollowed up like the bottom of a bottle ; under this the fire was placed : a copper tube attached to this bottom went around the inside of the boiler, and then passed out through the side of the boiler, the outside brick flues then carrying the heat around the outside of the boiler and into the chimney.

" Captain Trevithick's first plunger-pole lifts in Dolcoath were put in at this time and worked by this engine. Glanville, the mine carpenter, was head man over the engines when Captain Trevithick was away." [1]

" Charles Swaine worked as a rivet boy in making Captain Trevithick's cylindrical wrought-iron boilers for the Dolcoath engine. Several of Captain Trevithick's high-pressure boilers were working in the mines before that, but not made exactly like the Dolcoath engine boilers. When I was a boy about the year 1804, several years before I worked on the Dolcoath engine boilers, I carried father's dinner to the Dolcoath smiths' shop, where he worked, and used to stop and watch the wood beam going up and down of Captain Dick's first high-pressure steam-whim. She was not a puffer, but a puffer-whim worked near by, called the Valley puffer. At that time most of Captain Dick's high-pressure boilers were smallish, cast iron outside, and wrought-iron tube." [2]

In 1799, shortly after the reopening of Dolcoath Mine, Trevithick, jun., selected his father's second-hand atmospheric engine of 1775,[3] to further improve it by a new boiler of uniformly globular figure, with concave circular bottom, under which fire was placed ; it was of wrought iron, 24 feet in diameter, surrounded by external brick flues ; a large copper tube, starting from the boiler bottom, immediately over the fire,

---

[1] Henry Clark's recollections in 1869.
[2] Working in the Valley smiths' shop, in Dolcoath Mine, in 1869.
[3] See vol. i., p. 25.

served as an internal flue, carrying the fire by a sweep around the interior in the water space, and then out through the side of the boiler into the external brick flue. It may be said that there was nothing new in a circular form of boiler, or in an internal tube; but it will be admitted that this repaired engine, in this its third stride in the march of advancement, made publicly known those principles which in a few years more than doubled the power, the economy, and the applicability of the steam-engine. His patent drawing of 1802 shows this form of boiler applied to a small portable engine, in which, for the sake of simplicity of structure and cheapness, cast iron was used instead of wrought iron, and the internal tube omitted.[1]

The full detail estimate, from which the following items are extracted, of the cost of alteration was written by Trevithick, jun., in the book and on the page adjoining that containing the account of the former alteration and re-erection of the same engine by Trevithick, sen., in 1775.

"A 45-inch cylinder engine, working 20 lbs. to the inch :—

	£	s.	d.
Boilers, 8 tons at 42*l*.	336	0	0
Iron about ditto, 6 cwt. at 42*l*.	12	12	0
Castings about ditto, 15 cwt. at 42*s*.	18	0	0
Safety-valve and cocks	1	0	0
Wood about bob, 200 ft. at 6*s*.	60	0	0
Cast iron about ditto, 45 cwt. at 25*s*.	56	5	0
Brass about ditto, 60 lbs. at 2*s*.	6	0	0
Piston-rod, 4 in., 14 ft. long, 550 lbs. at 1*s*.	27	10	0
T-piece, 10 cwt. at 25*s*.	12	10	0
Cover, and bottom, and piston, 35 cwt. at 32*s*.	56	0	0
Nozzles, 6 cwt. at 32*s*.	9	12	0
Steam and perpendicular pipe, 10 cwt. at 25*s*.	12	10	0
Receiver, 2 ft. 4 in. long, and bottom, 15 cwt. at 25*s*.	18	15	0
Air-pump, bottom, and case, 10 cwt. at 25*s*.	12	10	0
Plunger, 22 in., 6 ft. long, 12 cwt. at 40*s*.	24	0	0
Force lift	5	0	0
Engineer	66	0	0"

---

[1] See vol. i., p. 128.

The term "single" refers to its open-top cylinder as originally erected by Newcomen, when it was called the Carloose engine, and so it remained after its re-erection in 1775, under the name Dolcoath new engine, alias Bullan Garden; but after the last re-erection in 1799 it had a cylinder-cover, and was called the Shammal 45-inch engine; "working 20 lbs. to the inch" meant the force on each inch of the piston, including vacuum on the one side of 14 lbs and steam on the other side of 6 lbs. to the inch.

Watt, on his first visit to Cornwall, in 1777, spoke disparagingly of the Newcomen atmospheric engines "burdened to 6 or 7 lbs. net to the inch." Fifty years later Stuart described Watt's engine as "using steam of a somewhat higher temperature than 212 degrees, so as to produce a pressure between 17 and 18 lbs. on each square inch of the piston; yet in practice, from imperfect vacuum and friction, it cannot raise more water per inch than would weigh about $8\frac{1}{2}$ lbs.,"[1] or an increase of net force—when compared with the Newcomen atmospheric—of only a pound or two on the inch in the lapse of years embracing the active lifetime of Watt. The cause of this slight increase of power is so simple that it has been passed by unnoticed by very many. The steam pressure in the Newcomen atmospheric was continued unaltered in the Watt vacuum engine. Trevithick constructed the first boiler and engine capable of safely and economically using the power of high-pressure steam. Nelson was obliged to come to close quarters, that his shot, propelled by weak cannon and low-pressure powder, might penetrate wooden ships. We now manufacture and control high-

---

[1] Stuart's 'History of the Steam-Engine,' published 1824.

pressure powder, so that 12 inches of iron armour-
plates cannot resist its force ; but this knowledge has
taken nearly as long in growing to perfection as did
the mastery of high-pressure steam, and its use in the
much more complicated steam-engine.

Watt's engine, as described a quarter of a century
after the expiration of his patent and the advent of the
high-pressure steam-engine, still derived its gross force
from 14 lbs. of vacuum and 2 or 3 lbs. of steam, resulting in a net force of 8½ lbs. Trevithick's engine
of 1799, which heralded the last hours of the Watt
patent authority, and may be taken as the first distinct
evidence of comparatively high-pressure steam in large
Cornish pumping engines, derived its power from 14 lbs.
of vacuum and 6 lbs. of steam, being together but 2 or
3 lbs. on the inch more than the Watt engine, but its
net force of 12 lbs. to the inch was half again as much
as the net force of the Watt engine, the increase being
wholly from the steam pressure, which was never practised by Watt, and which in its almost unlimited force
gives the greatly increased power to modern steam-
engines.

Trevithick's estimate for a new engine of the same
size as the old was 2000*l*., but as the old one could be
improved for 1300*l*., the latter course was adopted,
the wooden main beam with its segment head was retained, a cover was added to the cylinder, and a new
piston-rod and piston; a pole air-pump was used in
lieu of the more usual Watt air-pump bucket; a feed-
pole forced water into the boiler,—an indirect proof
of increased steam pressure. The new globular boiler
with internal tube weighed 8 tons; the engineer's charge
for carrying out the work was 66*l*.

The use of strong steam as the prime mover of the

steam-engine increased more rapidly beyond than within the limits of Cornwall, for in 1802 was erected at Coalbrookdale a high-pressure steam-puffer engine, to which Trevithick attached a pump which forced water through a column of upright pipes, that the power of the engine might be accurately measured. It worked with steam of from 50 to 145 lbs. on the inch, and wholly discarded the vacuum which had been Watt's mainstay.

"The boiler is 4 feet diameter, the cylinder 7 inches diameter, 3-feet stroke. The water-piston is 10 inches in diameter, drawing and forcing 35 feet perpendicular, equal beam. I first set it off with about 50 lbs. on the inch pressure against the steam-valve, for the inspection of the engineers about this neighbourhood. The steam continued to rise the whole of the time it worked; it went from 50 to 145 lbs. to the inch.

"The engineers at this place all said that it was impossible for so small a cylinder to lift water to the top of the pumps, and degraded the principle, though at the same time they spoke highly in favour of the simple and well-contrived engine.

"After they had seen the water at the pump-head, they said that it was possible, but that the boiler would not maintain its steam at that pressure for five minutes; but after a short time they went off, with a solid countenance and a silent tongue."[1]

This high-pressure steam pumping engine in 1802 may be taken as the first pumping engine of the puffer class using such strong steam.

In the spring of the following year[2] a somewhat similar engine was erected in London. "The cylinder is 11 inches in diameter, with a 3½-feet stroke. It requires the steam at a pressure of 40 to 45 lbs. to the inch to do its work well, working about twenty-six or

---

[1] See Trevithick's letter, August 22nd, 1802, vol. i., p. 153.
[2] See Trevithick's letter, May 2nd, 1803, vol. i., p. 158.

twenty-seven strokes per minute. It is much admired by everyone that has seen it, and saves a considerable quantity of coal when compared with a Boulton and Watt. Mr. Williams, Mr. Robert Fox, Mr. Gould, and Captain William Davey were here, and much liked the engine; they gave me an order for one for Cornwall as a specimen." This particular engine was for driving machinery in a cannon manufactory. A high-pressure pumping engine was at work at Greenwich, and some were at work in Cornwall.

"PENYDARRAN, near CARDIFF,
"MR. GIDDY, "October 1st, 1803.

"Sir,—In consequence of the engine bursting at Greenwich, I have been on the spot to inspect its effects. I found it had burst in every direction. The bottom stood whole on its seating; it parted at the level of the chimney. The boiler was cast iron, about 1 inch thick, but some parts were nearly $1\frac{1}{2}$ inch; it was a round boiler, 6 feet diameter; the cylinder was 8 inches diameter, working double; the bucket was 18 inches diameter, 21 feet column, working single, from which you can judge the pressure required to work this engine. The pressure, it appears, when the engine burst, must have been very great, for there was one piece of the boiler, about 1 inch thick and about 5 cwt., thrown upwards of 125 yards; and from the hole it cut in the ground on its fall, it must have been nearly perpendicular and from a very great height, for the hole it cut was from 12 to 18 inches deep. Some of the bricks were thrown 200 yards, and not two bricks were left fast to each other, either in the stack or round the boiler. It appears the boy that had care of the engine was gone to catch eels in the foundation of the building, and had left the care of it to one of the labourers; this man, seeing the engine working much faster than usual, stopped it, without taking off a spanner which fastened down the steam-lever, and a short time after being idle it burst, killed three on the spot and another died soon after of his injuries. The boy returned that instant, and was then going to take off the

trig from the valve.   He was hurt, but is now recovering; he
had left the engine about an hour.   I would be much obliged
to you if you would calculate the pressure required to burst this
boiler at 1 inch thick, supposing it to be a sound casting, and
what pressure it would require to throw the materials the distance I have before stated, for Boulton and Watt have sent a
letter to a gentleman of this place, who is about to erect some
of those engines, saying that they knew the effects of strong
steam long since, and should have erected them, but knew the
risk was too great to be left to careless enginemen, and that it
was an invention of Mr. Watt, and the patent was not worth
anything.   This letter has much encouraged the gentlemen of
this neighbourhood respecting its utility; and as to the risk
of bursting, they say it can be made quite secure.   I believe
that Messrs. Boulton and Watt are about to do me every injury
in their power, for they have done their utmost to report the
explosion, both in the newspapers and in private letters, very
different to what it really was; they also state that driving a
carriage was their invention; that their agent, Murdoch, had
made one in Cornwall and shown it to Captain Andrew Vivian,
from which I have been enabled to do what I have done.   I
would thank you for any information that you might have
collected from Boulton and Watt, or from any of their agents,
respecting their even working with strong steam, and if
Mr. Watt has ever stated in any of his publications the effects
of it, because if he condemns it in any of his writings, it will
clearly show from that, that he did not know the use of it.
Mr. Homfray, of this place, has taken me by the hand, and will
carry both the engines and the patent to the test.   There are
several of Boulton and Watt's engines being taken down here,
and the new engines being erected in their place.   Above 700
horse-powers have been ordered at 12l. 12s. for each horse-power
for the patent right, and the persons that ordered them make
them themselves, without any expense to me whatever.   If I
can be left quiet a short time I shall do well, for the engines
will far exceed those of Boulton and Watt.   The engine at
Greenwich did fourteen millions with a bushel of coals; it was
only an 8-inch cylinder, and worked without an expansive cock,

and under too light a load to do good duty; also on a bad construction, for the fly-wheel was loaded on one side, so as to divide the power of the double engine, and connected to the pump-rods on a very bad plan. I remember that Boulton and Watt's 20-inch cylinders when on trial did not exceed ten millions; I believe you have the figures in your keeping. Let us have the 60-horse power at work that is now building, and then I will show what is to be done. It will be loaded at 30 lbs. to the inch on each side the piston, it has an 8-feet stroke with an expansive cock, and the blowing cylinder directly over the steam-cylinder, as free from friction as possible. There was no engine stopped on account of this accident; but I shall never let the fire come in contact again with the cast iron. The boiler at Greenwich was heated red hot and burnt all the joints the Sunday before the explosion.

" I have received a letter from a person in Staffordshire who has a cylinder-boiler at work with the fire in it, and he says the engine performs above all expectation; he requests me to give him leave to build a great many more. I shall put two steam-valves and a steam-gauge in future, so that the quicksilver shall blow out in case the valve should stick, and all the steam be discharged through the gauge. A small hole will discharge a great quantity of steam at that pressure. There will be a rail-road-engine at work here in a fortnight; it will go on rails not exceeding an elevation of one-fiftieth part of a perpendicular and of considerable length. The cylinder is $8\frac{1}{2}$ inches in diameter, to go about two and a half miles an hour; it is to have the same velocity of the piston-rod. It will weigh, water and all complete, within 5 tons.

" I have desired Captain A. Vivian to wait on you to give you every information respecting Murdoch carriage, whether the large one at Mr. Budge's foundry was to be a condensing engine or not.

" Is it possible that this engine might be burst by gas?

" I am, Sir,

" Your very obedient servant,

" RICHARD TREVITHICK."

This high-pressure puffer pumping engine at Greenwich, in 1803, worked a pump of 18 inches in diameter. The engine boy having fixed the safety-valve while he fished for eels, caused an explosion of the boiler. This was the first mishap from the use of high-pressure steam. The boiler was globular, 6 feet in diameter, and from an inch to an inch and half in thickness, made of cast iron; the cylinder, of 8 inches in diameter, was partly let into and fixed on the boiler. Its general design is seen in the patent drawing of 1802, Fig. 1.[1] Trevithick determined in future to use two safety-valves, and also a safety steam-gauge. At that time one of his high-pressure puffer-engines, with a cylindrical boiler and internal tube, was working in Staffordshire.

The Greenwich high-pressure puffer-engine did fourteen millions of duty with a bushel of coals, 84 lbs. A 60-horse-power engine was being built in Wales, with an 8-feet stroke, to work expansively with 30 lbs. of steam on the inch in the boiler. For a more thorough test with the low-pressure vacuum engines, in competition, the Government intended to use the new engines, and some of Watt's engines having been removed to make room for them, Boulton and Watt wrote to a gentleman who was about to order an engine from Trevithick, " We knew the effects of strong steam long since, and should have erected them, but knew the risk was too great." Moreover, " it was an invention of Mr. Watt's, and the patent (Trevithick's) was not worth anything." This admission clearly shows not only that Watt did not make high-pressure steam-engines, but that he did his best to prevent others from making them.

---

[1] See vol. i., p. 128.

"MR. GIDDY,   "PENYDARRAN, CARDIFF, *January 5th*, 1804.

"Sir,—I received yours a few days since, and should have answered it sooner, but I was at Swansea for the last four weeks, and wished to return here to give you as full an account of our proceedings as possible.

"We have had an 8-inch cylinder at work here by way of trial; it worked exceedingly well a hammer of the same size as is now being worked here by an atmospheric engine 28 inches diameter, 5-feet stroke, which does not master its work with greater ease than the 8-inch cylinder. The 8-inch is now removed to Swansea, and is winding coals; the baskets hold 6 cwt. of coal; it lifts 80 yards in a minute and a quarter, and burns 6 cwt. of coal in twenty-four hours. There were twelve horses on this pit before, lifting 80 tons of coal in the course of the twenty-four hours. You may fairly state that the 8-inch cylinder does between thirty and forty horses' work in twenty-four hours, with 6 cwt. of coal.

"One of Boulton and Watt's 18-inch double engine, about half a mile from it, lifting baskets of the same size, and with the same velocity, burns above three times the quantity of coal.

"The 8-inch engine requires the steam to be about 46 or 48 lbs. to the inch to do its work well. The standers-by would not believe that such a small engine could lift a basket of coal, but are now much pleased with it, and have given orders for several more. There will be another at work here for the same purpose in about six weeks, a 15-inch cylinder, 6-feet stroke, which is a great power for a winding engine.

"Mr. Watt says, in a letter to Mr. Homfray, that he could not make any of his experiments in strong steam answer the purpose. It is my belief that he never made any experiments of any consequence in strong steam.

"A great number are building at different foundries. Mr. Sharratt, a founder at Manchester, who has four in building, said that he would not pay the patent right; on giving him notice of a trial he agreed to pay the patent right.

"I have received a letter from London, saying that an engineer called Dixon has two engines on the same plan working; and says that he shall not pay anything to the patentee; that the words in Mr. Watt's specification are enough to indemnify him from my threats. We have had three counsels' opinions on the subject, and they all agree that the patent is good. Counsels Marratt and Gibbs principally treated on the construction of the engine, more than on the principle; but Erskine was principally on the principle of the engine, and said very little of its construction. They all say the words in Mr. Watt's specification will have no weight whatever against us.

"I shall leave this place to-morrow for London to make inquiry into those engines, and to get the business into court if they will contend. I shall be at No. 2, Southampton Street, Strand, and expect to be in town about five or six days, and if you will be so good as to return here, from Oxford, with me, I will call on you in my journey down. It is but 50 miles from Bristol, and not so much as 100 miles from Oxford, and the coach passes very near this place.

"There is a great deal of machinery and mining here, which would engage your attention for a few days, and very pleasant gentlemen about the neighbourhood.

"If I had not been called to Swansea to put up the winding engine, the road-engine would have been at work long since, but in my absence very little was done to it. The work is all ready, and a part of it put together. If I could tarry four or five days longer I could set it to work before going to London. They promise me that it shall be completed before my return. I think there is no doubt of its being finished, as I have Frank Bennetts here from Cornwall about it, and a plenty of hands to assist him.

"I have a thousand things to relate to you, too much for paper to contain, therefore must request you to be so good as to go down from Oxford with me, and I will promise, on warrant, that the road-engine shall be finished before my return. When it is set to work I shall return to Cornwall.

"I remain, Sir,

"Your humble servant,

"RICHARD TREVITHICK."

In 1804 an 8-inch cylinder high-pressure puffer-engine, with steam of 48 lbs. to the inch, worked a large hammer as well as a 28-inch cylinder atmospheric engine, and more economically than a Watt low-pressure steam vacuum engine with an 18-inch cylinder, which was five times as large as the little high-pressure. In consequence of this superiority those who came to witness the trial ordered several more of Trevithick's engines, one of which with a 15-inch cylinder and 6-feet stroke was to be at work in a few weeks.

Watt wrote to Mr. Homfray " that he could not make any of his experiments in strong steam answer the purpose," and Trevithick declared Watt never could have tried any experiments with high steam.

Dixon refused to pay patent right because the words of Mr. Watt's specification, "in cases where cold water cannot be had in plenty, the engines may be wrought by the force of steam only, by discharging the steam into the open air after it has done its office," " are enough to indemnify him." Eminent counsel were of opinion that " the words in Watt's specification will have no weight whatever."

Marratt and Gibbs were inclined to rest on the difference in the construction of the two kinds of engines, while Erskine boldly said that the principle was different, and he cared little for the kind of construction.

The admission by Watt that he could do nothing with high steam after an experience of thirty years from the date of his patent, shows how difficult the work was to those who had to find the way; yet Trevithick had several at work within a few months of his first mental sight of a steam-engine without condensing water, fitful glimpses of which passed and repassed

while he sat unobserved in the crowded law court in
1796 hearing the remarks of engineers and counsel.

" The public until now called me a scheming fellow, but
their tone is much altered. An engine is ordered for the West
India Docks, to travel itself from ship to ship, to unload and to
take up the goods to the upper floors of the storehouses.

"Boulton and Watt have strained every nerve to get a Bill
in the House to stop these engines, saying the lives of the public
are endangered by them, and I have no doubt they would have
carried their point, if Mr. Homfray had not gone to London to
prevent it ; in consequence of which an engineer from Woolwich was ordered down, and one from the Admiralty Office,
to inspect and make trial of the strength of the materials." [1]

After a week or two another letter states,[2]—

" We are preparing to get the materials ready for the experiments by the London engineers, who are to be here on Sunday
next. We have fixed up 28 feet of 18-inch pumps for the
engine to lift water.

" These engineers particularly requested that they might
have a given weight lifted, so as to be able to calculate the
real duty done by a bushel of coal.

" As they intend to make trial of the duty performed by the
coal consumed, they will state it as against the duty performed
by Boulton's great engines, which did upward of twenty-five
millions, when their 20-inch cylinders, after being put in the
best order possible, did not exceed ten millions. As you were
consulted on all those trials of Boulton's engines, your presence
would have great weight with those gents, otherwise I shall not
have fair play.   Let me meet them on fair grounds and I will
soon convince them of the superiority of the ' Pressure-of-steam
engine.' "

Watt left no stone unturned to prevent the use of
high-pressure steam-engines, and fortune favoured

---

[1] See Trevithick's letter, 22nd February, 1804, vol. i., p. 161.
[2] See Trevithick's letter, 4th March, 1804, vol. i., p. 166.

him, for after four or five days Trevithick again wrote :—

"I am sorry to inform you that the experiments that were to be exhibited before the London gents are put off, on account of an accident which happened to Mr. Homfray. I find myself much disappointed on account of the accident, for I was desirous to make the engine go through its different work, that its effect might be published as early as possible."[1]

While constructing those numerous high-pressure engines for rolling mills, winding engines, and pumping engines, the Welsh and Newcastle locomotives were being made and worked, yet he found time to teach the people of Stourbridge.

"MR. GIDDY,   "STOURBRIDGE, *July 5th*, 1804.

"Sir,—I should have answered your letter some time since, but waited to set two other engines to work first. The great engine at Penydarran goes on exceedingly well. The engine will roll 150 tons of iron a week with 18 tons of coal. The two engines of Boulton's at Dowlais burn 40 tons to roll 160 tons; they are a 24-inch and a 27-inch double. The engine at Penydarran is 18½ inches, 6-feet stroke, works about eighteen strokes per minute: it requires the steam about 45 lbs. to the inch above the atmosphere. I worked it expansive first, when working the hammer, which was a more regular load than rolling; then with steam high enough to work twelve strokes per minute with the cock open all the stroke; then I shut it off at half the stroke, which reduced the number of strokes to ten and a half per minute, the steam and load the same in both; but I did not continue to work it expansively, because the work in rolling is very uneven, and the careless workmen would stop the engine when working expansive.

"When the cylinder was full of steam the rollers could not stop it; and as coal is not an object here, Mr. Homfray wished

---

[1] See Trevithick's letter, 9th March, 1804, vol. i., p. 168.

the engine might be worked to its full power. The saving of coal would be very great by working expansively.

"The trials we have made for several weeks past against Boulton's engines have been by working with the cylinder full of steam. The cock springs out of its seat when water gets into the cylinder, and prevents any mischief from the velocity of the fly-wheel.

"The tram-engine has carried two loads of 10 tons of iron to the shipping place since you left this. Mr. Hill says he will not pay the bet, because there were some of the tram-plates in the tunnel removed so as to get the road into the middle of the arch.

"The first objection he started was that one man should go with the engine, without any assistance, which I performed myself without help; and now his objection is that the road is not in the same place as when the bet was made.

"I expect Mr. Homfray will be forced to take steps that will force him to pay. As soon as I return from here there will be another trial, and some person will be called to testify its effects, and then I expect there will be a lawsuit immediately. The travelling engine is now working a hammer.

"At Worcester last week we put a 10-horse engine to work in a glover's manufactory. The flue from the engine is carried through the drying room and dries his leather. The steam from the engine goes to take the essence out of the bark, and also to extract the colour out of the wood for dyeing the leather. Then it boils the dye, and the steam that is left is carried into his hot-house. It works exceedingly well. This week I put another to wind coals at this place, a 10-horse power, which works very well. All the tradesmen are set against it; they say that there is no carpenter or mason work about it, and very little smith-work, and that it will destroy their business. The engineer on the spot is also against it very much. I do not expect that it will be kept long at work after I leave it, unless the proprietor takes care to prevent those people from doing an injury to it. Mr. Homfray was here yesterday, but is now returned to Peny-darran. I shall go from here to Coalbrookdale.

"There is an engine there almost ready for the West India

Docks. It will be ready to send off to London in about four weeks. It will be a very complete engine. The pumps for forcing the water will be fixed on the back of the boiler. It will force 500 gallons of water 100 feet high in a minute ; above ten times the quantity that engines worked by men can do. Mr. Homfray and myself shall be in town as soon as the castings are sent off. I hope you will be there at the time. If you wish to see the engines already at work in London, call on Mr. David Watson, steam-engine maker, Blackfriars Road. He lives up about 500 or 600 yards above the bridge on the left-hand side; you will see his name over his door. If you have time to inspect those engines you will find by comparing them against Boulton's, doing the same work, that there is a great saving of coal above other engines. . . . I shall go to Liverpool and Manchester from here, and again to Coal-brookdale.

"There are three engines at the Dale begun, to work with condensers, for places where coal is scarce. I think it is better to make them ourselves, for if we do not, some others will, for there must be a saving of coal by condensing. But with small engines, or where coal is plentiful, the engine would be best without it. They say at the Dale about putting two cylinders, but I think one cylinder partly filled with steam would do equally as well as two cylinders.

"That engine at Worcester shuts off the steam at the first third of the stroke, and works very uniformly. I cannot tell what coal it burns yet, but I believe it is a very small quantity. I shall know in a short time what advantage will be gained by working expansive. I expect it will be very considerable. There are a great many engines making and ordered. Boulton and Watt and several others are doing everything to destroy their credit, but it is impossible to destroy it now that it is so well known. I have not taken any of the ground at Bristol to remove. I called on them and told them it was possible to break the ground without men, and they wish me to take a piece to clear out, but would not set but a small piece at a time; therefore it would be disclosing the business to no purpose. They were very desirous to know the plan, but I would

not satisfy them, neither will I unless they pay me for it in
some way or other. If you direct for me at the Dale it will
find me. I am happy to find that you have a seat in the House.
I wish every seat was filled with such.

"I remain, Sir,

"Your very humble servant,

"RICHARD TREVITHICK."

Trevithick fully understood the value of the expansive principle in 1804 : when working with steam of
45 lbs. to the inch, the engine went at a speed of
twelve strokes a minute. On cutting off the steam at
half-stroke, the speed and consequent work done fell
to ten and a half strokes a minute; in other words, the
work performed by the engine fell off only one-eighth
part, while the quantity of steam and consequently of
coal was reduced by one-half. The principle was established, but the application was practically incomplete
from the want of heavier fly-wheels, to give out their
momentum during the latter half of the stroke, when
the expanding steam was lessening its force.

"The saving of coal would be very great by working
expansively, but as coal is not an object here," Mr.
Homfray was careless about the expansion. Thirty-
three years after this indirect check to steam-engine
economy, the writer, then living in the Sirhowey Iron
Works, and within stone's-throw of Mr. Homfray's
Works, recommended the removal of the Boulton and
Watt's waggon boilers, to make room for Trevithick's
boilers, on the plea of saving one-half the fuel, and at
the same time increasing the power of the engine, and
thereby the pressure of the blast in the iron furnaces.
The proprietor was careless about the saving of coal,
and was doubtful that an increased blast would increase

the quantity of iron smelted. The promise that the
wages of one-half of the number of boiler firemen would
be saved, was understood. Trevithick's high-pressure
boilers replaced the Watt low-pressure, resulting in a
largely - increased quantity of iron from the greater
power and pressure of blast in the furnaces, and at one-
half the expenditure of coal in the boilers: ten men
had been employed as firemen of the Watt boilers
during twenty-four hours; with Trevithick's boilers,
five men did the work.

The high-pressure puffer-engine, with an 18-inch
cylinder, working with 45 lbs. of steam, rolled as much
iron as the two larger low-pressure vacuum engines of
Watt, of 24 and 27 inch cylinders, which together were
more than three times the size of the high-pressure
engine, and cost three times as much.

At Stourbridge, as elsewhere, everyone was against
the new plan. The engineer in charge did not like it,
and the carpenters, smiths, and masons saw the end
of their occupation as engine erectors, if there was no
longer a necessity for foundations, well-work, &c., for
condensing water, and many other things, necessary
to complete a Watt engine; while the high-pressure
puffer was no sooner unloaded than it was ready to
work.

A great charm in Trevithick's character was his
freedom and largeness of view in questions of competition. He was then making three engines at Coalbrookdale, to be worked with high-pressure steam, combined
with the Watt air-pump and condenser; and though
smarting from the contest with his great rival, yet
wrote, " I think it is better to make them ourselves, for
if we do not, some others will, for there must be a
saving of coal by condensing. But with small engines,

or where coal is plentiful, the engine would be best
without it."

Those words accurately describe the practice of the
present day, though written sixty-six years ago, and
were followed by others equally true in principle,
though varied in form to suit special requirement.
" They say at the Dale about putting two cylinders,
but I think one cylinder partly filled with steam would
do equally as well as two cylinders."

These sagacious views required the untiring labour
of the following twelve years to perfect and make practical, when applied to the largest engines of the time;
which we shall now trace in the construction of a strong
and economical boiler, supplying high-pressure steam to
the cylinder during only a comparatively small portion
of the stroke, completing it by expansion, so that at
its finish the steam had become of low pressure when
passed to the condenser. The moving parts and expansive gear were so simplified as to be applicable to the
then existing low-pressure steam vacuum engines without the complication of the double cylinders of Hornblower and Woolf.

"DEAR SIR,   " PENYDARRAN PLACE, *December 26th*, 1804.

"I have been favoured with your letter, and in answer,
respecting Mr. Mitchell, I am at a loss to know from your letter
what kind of iron he may likely want. If you will direct him
to write to me, and explain himself, I will immediately reply to
him and do what I can to assist and serve him. I believe there
are vessels going over frequently from Cardiff to Cornwall with
coals, that he might have part in cargo and the remainder in
coals. I am happy to give you the most satisfactory account of
our ' Trevithick's engine' going on well. It has now been at
work many months, and is by far the best engine we have.
We have for weeks weighed the coal, and knowing the work it

does, can speak with confidence. Its 18 inches diameter steam-cylinder consumes as near as can be 3 tons of coal in twenty-four hours, or 18 tons per week; and in this time it rolls with ease 130 tons long weight of iron from the puddling furnaces, at the same heat, into bars of 3 inches by about half an inch thick. Now, one on Messrs. Boulton and Watt's plan, of '24 inches' steam-cylinder, at our neighbouring works at Dowlais, employed in doing exactly the same kind of work, consumes *full* as much coal, and rolls only 90 tons in the week. These being facts, open for any person daily to see, must convince any dispassionate man of the superiority of 'Trevithick's engines,' and that the saving of fuel is nearly one-third, besides the other advantages of saving water and grease, which is no little. The packing of the piston now gives us little or no trouble, it goes from a fortnight to a month, opening the top now and then to screw it down, as it gets slack, which should be attended to. We use no grease or oil in packing the piston or working the engine, having found blacklead mixed with water, and poured 'a little now and then' through a hole on the top into the steam-cylinder, suits the packing of the piston much better, and is cheaper than anything else. About 1s. worth of black-lead will last our engine a week. We are now so thoroughly convinced of the superiority of these engines that I have just begun another of larger size. The boiler is to be 24 or 26 feet long, 7 feet diameter, fire-tube at wide end 4 feet 4 inches, and at narrow end, where it takes the chimney, 21 inches, steam-cylinder 23 inches diameter. This boiler, on account of the length of its tube withinside, will, I have no doubt, get steam in proportion, and work the engine with much less coals than our present one. Trevithick is at Coalbrookdale, Manchester, &c., &c., very busy, a great number of engines being in hand in that part of the world; and I think by perseverance the prejudice is wearing away very fast, and in spite of all Messrs. Boulton and Watt's opposition, they must and will take the lead of theirs. Any person now wanting engines, must be next kin to an idiot to erect one of Boulton's in preference to Trevithick's. I find there is a small one making near you by Mr. Vivian. I hope they have corresponded with Trevithick about the proportions of it; if they have not, I shall be particularly obliged to you to desire them to do so, for by his experience of what he has done they may be benefited, for it would be a shocking thing to have a bad engine put up for the first time in his native county.

"Mrs. Homfray unites with me in best compliments, and wishing you many happy returns of the season.

"I remain, dear Sir,

"Your most obedient servant.

"SAMUEL HOMFRAY.

"To MR. DAVIES GIDDY."

The evidence in this contest between the Watt low-pressure steam vacuum engine and the Trevithick high-pressure steam-puffer engine is in favour of the new principle; for the steam-engine with an 18-inch cylinder did fifty per cent. more work than the vacuum engine with a 24-inch cylinder with an equal quantity of coal, though the latter was seventy-five per cent. larger than the former; and a still greater economy was expected from the larger boiler to be built, 26 feet long, 7 feet in diameter, with internal fire-tube 4 feet 4 inches diameter at the fire end, tapering to 21 inches at the chimney end.

Thus in 1804 the cylindrical boiler in Wales had nearly reached its present form, and Homfray thought that none but idiots would prefer the Watt engine; forgetting that Trevithick's near friends and neighbours were carrying on a similar contest at Dolcoath Mine.

"PENYDARRAN PLACE, *January 2nd*, 1805.

"MR. DAVIES GIDDY,

"Dear Sir,—I have duly received your favour enclosing a letter for Mr. Trevithick, and which I, according to your desire, forwarded to him at Manchester, where he now is; and a

letter directed to him, to the care of Mr. Whitehead, Soho
Foundry, Manchester, will find him, as he will stay a little time
there, being very busy. I had lately the pleasure of writing to
you, and gave you the account of our engine working, and the
satisfaction it gives; I have nothing more to add on the subject, but that it is now at work, going on as usual, and I should
be happy for you to have a sight of it.

"We are beginning another of a larger size, and I have no
doubt but by making the cylindrical boiler larger, so as to take
a longer tube withinside it, by which means the fire will spend
itself before it leaves the tube to go up the chimney, that we
shall work to much better advantage in point of fuel than we
do at this present one, as this boiler is so short that a great deal
of the flame of the fire goes up the chimney. We are now
better acquainted with the different proportions than we at first
were, for which reason I am anxious that one now making by
Mr. Vivian should be made according to the directions of
Mr. Trevithick.

"I beg leave to offer you the compliments of the season, and
many happy returns, and

"Remain, respectfully, dear Sir,

"Your most obedient servant,

"SAMUEL HOMFRAY."

Trevithick, always busy, was just now doing the
work of a host, for everybody had to be taught how
to make high-pressure steam-engines; and the Newcastle locomotive, the Thames steam-dredging, and
other special applications of steam-power required his
presence, especially the fight with Watt at Dolcoath
Mine, where Andrew Vivian, as mine manager, was
erecting a high-pressure steam-puffer whim-engine to
compete with a Watt low-pressure steam vacuum whim-
engine.

"The adventurers grumbled because Captain Trevithick was
so often away from the mine. Glanville, the mine carpenter,

the head man over the engines, made a trial between Trevithick's high-pressure puffer whim and Watt's low-pressure condenser. When Captain Trevithick heard of it, he wrote down from London that he would bet Glanville 50l. that his high-pressure puffer should beat Watt's low-pressure condenser. Then he came down from London and found that the piston of his engine was half an inch smaller in diameter than the cylinder. When a new piston was put in, she beat Boulton and Watt all to nothing. Persons were chosen to make a three or four weeks' trial, and when it was over, 'a little pit was found with coal buried in it, that Glanville meant to use in the Watt engine.'"[1]

Pooly, Smith, and others, say that Trevithick's Dolcoath puffer had the outer case of the boiler of cast iron, the fire-tube of wrought iron, the cylinder horizontal, and fixed in the boiler. Captain Joseph Vivian saw Trevithick's whim in Stray Park Mine about 1800 or 1801, and a similar one was erected in Dolcoath, and after a year or two a Boulton and Watt low-pressure whim was put up to beat it. The trial was in favour of the Watt engine, but everybody said the agents were told beforehand which way the report ought to go; so the engine that *puffed the steam up the chimney* was beaten.

Trevithick, who was busily engaged in Manchester at that time, the early part of 1805, when informed of what was going on in Cornwall, wrote:—

" I fear that engine at Dolcoath will be a bad one. I never knew anything about its being built until you wrote to me about Penberthy Crofts engine, when you mentioned it. I then requested Captain A. Vivian to inform me the particulars about it, and I find that it will not be a good job. I wish it never was begun."[2]

---

[1] Recollections of Henry Clark, living at Redruth in 1869.
[2] See Trevithick's letter, January 10th, 1805, vol. i., p. 324.

" Mr. Giddy,   "Camborne, *February* 18*th*, 1806.

" Sir,—On my return from town I altered the pressure
of the steam-engine at the bottom of the hill, Dolcoath. Before
I returned there was a trial between mine and one of Boulton's;
both engines in the same mine and drawing ores from the same
depth. The result was, Boulton's beat the pressure-engine as
120 to 55. Since it was altered there have been three other
trials; the result was 147 to 35 in favour of the pressure of the
steam-engine. They are now on trial for another month, and at
the next account *they intend to order a new boiler for the great
engine,* and work with high-pressure steam and condenser, provided this engine continues to do the same duty as was done
in the former trials. This engine is now drawing from a perpendicular shaft, and Boulton and Watt's from an underlay
shaft; but to convince Captain Jos. Vivian, we put it to draw
out of the worst shaft in the mine, and then we beat more than
three to one; we lifted in forty-seven hours, 233 tons of stuff
100 fathoms with 47 bushels of coal. The engine was on trial
sixty-six hours, but nineteen hours were hindered by the shaft
and ropes, &c., which made the consumption of coals about ¾ths
of a bushel per hour. The fire-tube is 2 feet 3 inches diameter,
and the fire-bars were only 14 inches long. The fire-place was
but 2 feet 3 inches wide by 14 inches long; and the fire about
4 or 5 inches thick; it raised steam in plenty; it was as bright
as a star. The engine is now doing the work of two steam-
whims; the other steam-whim in the Valley is turned idle,
and both shafts will not more than half supply it. 233 tons
are equal to nearly 2000 kibbals, which were drawn in forty-
seven hours.

" Mr. Harris has a 12-inch cylinder making at Hayle, for
Crenver, and Mr. Daniel has a 14-inch for Perran-sand, and
a great number are waiting for the trial of this month, *before
altering their boilers to the great engines.*

" The steam-whim that is now turned idle at the Valley was
13½-inch cylinder, 4-feet stroke; it turned the whim one revolution to one stroke, and lifted the kibbal the same height at a
stroke as my engine did, and I think took the same number of
gallons of steam to lift a kibbal as mine did. Their steam was

not above 4 lbs. to the inch; *mine was near* 40 *lbs. to the inch;*
yet I raised my steam of near 40 lbs. with a third of the coals
by which they got theirs of 4 lbs. to the inch. This is what I
cannot account for, unless it is by getting the fire very small and
extremely hot. Another advantage I have is, that there is no
smoke that goes off from my fire to clog the fire sides of the
boiler, while the common boilers get soot half an inch thick, and
the mud falls on the bottom of the boiler, where the fire ought
to act; but in these new boilers the mud falls to the bottom,
where there is no fire, and both the inside and outside of the
tube are clean and exposed both to fire and water. This fire-
place of 14 inches was 5 feet long when I came down, and then
the coal did not do above one-seventh of the duty that it now
does.

" I would be very much obliged to you for your opinion on
what I have stated, and what *advantage you think the great
engine is likely to get from working with steam about 25 lbs. to
the inch, and shut off early in the stroke, so as to have the steam
about 4 lbs. to the inch when the piston is at the bottom. I think
this, with the advantage of the fire-place, will make a great
saving.*

" The present fire-place is 22 feet from fire-door to fire-door,
9 feet wide, and 7 feet thick in fire. There is not one-tenth of the
coals that are in the fire-place on fire at the same time; it will
hold 30 tons of coals at one time, and I think that a great deal
of coal is destroyed by a partial heat before it takes fire. A
boiler on the new plan will not cost more than two-thirds of the
old way, and will last double the time, and can be cleaned in
three hours. It requires twenty-four hours in the old way, and
we need to clean the boilers only one-fourth the number of
times.

" Though these trials have shown so fairly that it is a great
advantage, my old acquaintances are still striving with all their
might to destroy the use of it; but facts will soon silence
them.

" I am about to enter into a contract with the Trinity Board
for lifting up the ballast out of the bottom of the Thames for all
the shipping. The first quantity stated was 300,000 tons per

year, but now they state 500,000 tons per year. I am to do
nothing but wind up the chain for 6d. per ton, which is now done
by men. They never lift it above 25 feet high. A man will
now get up 10 tons for 7s. My engine at Dolcoath has lifted
above 100 tons that height with 1 bushel of coals. I have two
engines already finished for this purpose, and shall be in town
in about fifteen days to set them at work. They propose to
engage with me for twenty-one years. The outlines of the
contract they have sent me down, which I think is on very fair
terms. I would thank you for your answer before I leave this
county.

" I am, Sir,

" Your very humble servant,

" Rd. Trevithick."

In the trial at Dolcoath during his absence the high-
pressure steam-puffer whim was beaten by Watt's low-
pressure steam vacuum whim-engine as 55 to 120; but
having corrected some oversight in the puffer-engine,
it then beat Watt as 147 to 35. The trial was to be
continued for a month; and provided the superiority of
his whim-engine could be maintained, the adventurers
would allow him to apply his high-pressure boilers to
their large Boulton and Watt pumping engine. The
trial with the whim-engines was for the greatest number of kibbals of mineral raised to the surface by the
least consumption of coal. A dispute arose on the difference of the shafts, the one causing more friction to
the moving kibbal than the other, when Trevithick
agreed to take the worst shaft in the mine. On a trial
during sixty-six hours Watt's engine was beaten by
more than four times; and as Trevithick's engine did
the work that before required two engines, one of the
low-pressure steam Watt engines was removed that the
engine working with 40 lbs. on the inch might perform
the whole work.

" My fire-tube is 2 feet 3 inches in diameter, and the fire-bars only 14 inches long, and the fire only about 4 or 5 inches thick; it raised steam in plenty, and was as bright as a star." These words certainly imply the use of the blast-pipe, making the fire as bright as a star, and enabling the small boiler to give the required supply of steam. Several high-pressure puffer-engines had been ordered, and many persons were waiting the conclusion of the month's public trial to enable them to judge between the Watt and the Trevithick engine.

" MR. GIDDY,   "CAMBORNE, *March 4th*, 1806.

" Sir,—The day after I wrote to you the first letter, I received yours, and this day I have yours of the 1st instant.

" I am very much obliged to you for the figures you have sent me. I am convinced that the *pressure of steam will not hold good as theory points it out, because on expanding it will get colder, and of course lose a part of its expansive force after the steam-valve shuts.* I think there can be no risk in making this trial on Dolcoath great engine, as they intend to have a new boiler immediately, so as to prevent stopping to cleanse; and a boiler on this new plan can be made for one-third less expense than on the old plan, when you count the large boiler-house and ashes-pit, and brickwork round the boiler. It is not intended to alter any part of the engine or condenser, but only work with high steam from this new boiler; and if this boiler only performs as good duty as the old one, it will be a saving of near 300*l.* to them on the erection. *The vast matter this great engine has in motion will answer in part the use of a fly-wheel :* the whole of the matter in motion is near about 200 tons, at a velocity of about 160 feet a minute. This I know will not be sufficient; but it will be about equal to a fly-wheel of 20 feet diameter, 25 tons weight, twenty rounds per minute, if weight and velocity answer the same purpose.

'' Since Monday, the 18th February, being Dolcoath account-day, both engines have been on trial, and are to be continued until the next account, 17th instant. The engines are kept on

in the usual way, as at other times. Neither of the engines
have done so much duty as on the first trials, as they have not
been so strictly attended to. The average of the trial at this
time stands 26 cwt. for a bushel of coals to Boulton and Watt's
engine; mine, 83 cwt. for a bushel of coals.

"If I do not remain in Cornwall to attend next Dolcoath
account, I shall be in town about the 15th instant, otherwise
about the 20th instant. I shall call on you immediately on my
arrival. In this time I should be glad to hear from you again.
The Trinity business will answer exceedingly well; I have two
engines ready for that purpose to put to work on my arrival in
town.

"I am, Sir,

"Your very humble servant,

"Rd. Trevithick.

"P.S.—I would try the evaporation of water by both boilers,
but Boulton and Watt's engine is so pressed with work, and
being on the best part of the mine, they will not stop it a
moment. A boiler of 8 feet diameter and 30 feet long will
have as much fire-sides in the tube as there is now in Dolcoath
great boiler. The fire-tube in this boiler would be 5 feet
diameter, and a fire-place 6 feet long in it would be 30 feet of
fire-bars. In the whim-engines I find that a fire-place 14 inches
long and the tube 2 feet 3 inches diameter would, being forced,
burn 1 bushel per hour. At this rate the great tube would
burn near 12 bushels per hour, which is above the quantity
that the great engine boiler can consume, now at work. Small
tubes would have an advantage over large ones. Two boilers
would not cost much more than one large one, and be much
stronger."

The battle-ground of the fight between low and high
pressure from 1806 to 1812 had also served for the
personal encounter of Trevithick, sen., and Watt a
quarter of a century before, when the Dolcoath great
pumping engine was erected to compete with the two

earlier atmospherics; all three were still at work, over-looked by Carn Brea hill and castle, once the resort of Druid priests, whose sacrificial rites are still traced, by the hollows and channels for the blood of victims on the granite rocks.

CARN BREA CASTLE.   [W. J. Welch.]

" MR. GIDDY,   "CAMBORNE, *March* 21st, 1806.

"Sir,—The trial between the two engines ended last Monday, which was Dolcoath day. Boulton and Watt's engine, per average of trial, 1 ton 20 cwt. 2 qrs., with 1 bushel of coals; the other, 5 tons 11 cwt. 3 qrs., with 1 ditto, the same depth of shaft. The adventurers ordered the new castings that were made for another of Boulton and Watt's engines to be thrown aside, and another new engine of mine to be built immediately. The great boiler for the old engine is not yet ordered.

" I have received orders for nine engines within these four weeks, all for Cornwall. Two 12-inch cylinders, two 16-inch ditto, three 9-inch ditto, one 8-inch ditto, one 7-inch ditto. I expect one will be put to work next week at Wheal Abraham, for lifting water.

" This day I shall leave Cornwall for London. Shall stop two days in the neighbourhood of Tavistock, and take orders for three engines. As soon as I arrive in town I will call at your lodgings. I expect that the patent will be brought into court about the end of May. A person in Wales owes us about 600*l*. patent premium, and he says that the patent is not good. More particulars you shall have on my arrival.

" The railroad is going forward. I have the drawings in hand for the inclined plane.

" I am, Sir,

" Your very humble servant,

" RD. TREVITHICK."

The fact that expansion of steam caused reduction of heat was so evident to Trevithick that he ventured to doubt his friend's theory. The trials between the whim-engines having continued a fortnight, showed that the high-pressure steam-puffer had lifted 83 cwt., while the low-pressure steam vacuum only lifted 26 cwt. with the consumption of a bushel of coal. A suitable high-pressure boiler for the Watt low-pressure steam 63-inch pumping engine should be 30 feet long, 8 feet in diameter, with an internal fire-tube 5 feet in diameter; proportions approved of in the present day. The recommendation in 1806 to use small tubes may claim to be the first practical decision on the advantage of tubular boilers; and at the same time we read of the first hesitating step on the part of the public to use high-pressure steam in a Watt low-pressure engine,

which was still deferred for further consideration, even
with the limited pressure of 25 lbs. to an inch; so the
large Watt pumping engines were doomed for another
four or five years to struggle through their work with
low-pressure steam, though at that time Cook's Kitchen
high-pressure expansive condensing whim-engine had
been for years at work close by.

The shareholders professed to have fear of explosion;
but party-feeling and ignorance were the real causes of
opposition, for working men had no dread of the new
engines, while influential men leaned toward Watt's
old-fashioned plans.

This fear of Trevithick's expansive plans and high
steam is the more surprising, because at that time a
new boiler was required for the Watt 63-inch cylinder
pumping engine and Trevithick's cylindrical tubular
boiler could be made for one-third less cost than
the Watt waggon boiler, thus saving 300*l.*, and in
addition he promised to apply the higher pressure of
steam to the Watt engine without any change in its
parts or expenditure of money, and make it set in
motion at the commencement of the stroke the 200 tons
of pump-rods, the momentum of which would, with the
expansion of the steam, when shutting it off soon after
the first start in the movement of each stroke, carry it
through to the end; and he practically compares this
advantage from hoarded momentum in the pumping
engine with his experience of the fly-wheel of the rolling-
mill expansive engine in Wales.

The whim-engine with a fire-tube 2 feet 3 inches in
diameter used 84 lbs. of coal per hour; and at that rate
one cylindrical boiler 30 feet long, 8 feet in diameter,
with internal fire-tube 5 feet in diameter, would supply
steam for Watt's 63-inch cylinder; but in place of it

he preferred two smaller boilers, because small tubes have an advantage over large ones, and are much stronger.

The whim trials—high-pressure puffer against low-pressure vacuum—went on for another fortnight, when high pressure, having done twice as much work as low pressure, with an equal consumption of coal, the adventurers threw aside the work that had been made for another Watt engine, ordering one in its stead from Trevithick; but they could not just then make up their minds to place the Watt 63-inch pumping engine in his hands.

" DEAR SIR,   " CAMBORNE, *May 30th*, 1806.

"I am very happy to find you have so far continued your agreement with the Trinity gents, and think the bargain is a good one.   Must still beg leave to remind you not to proceed to show what your engine will do till the agreement is fully drawn up and regularly signed.

"Dolcoath agents, since they are informed of the accident at the iron-works in Wales, *of the engine blowing to pieces,* have requested me to have your opinion whether the old cylinder is strong enough for the boiler of the intended new engine, or whether you would recommend them to have a new one.   Your answer to this as soon as possible, as Mr. Williams and some others are likely to make some objections.

"Mr. Sims, the engineer, has published in the Truro paper, that one of Boulton and Watt's engines at Wheal Jewell has drawn more than a ton of ore over and above that drawn by the Dolcoath engine from the same depth by a bushel of coal. On inquiry I found they had only tried for twenty-two hours. They said they left off with as good a fire as they began with. This I argued was not a fair trial.   They say they are now on a trial for a month.

" The little engine at Wheal Abraham does its duty extremely well.   The particulars as to consumption of coal cannot be fairly ascertained, as she has never been covered, is fed with cold water, and has not water to draw to keep her constantly at work.

"I wish I could give a better account of the mines than is in my power to give, or of the standard price for ore, though the latter is rather looking up than otherwise. Our friend, North Binner Downs, is better than paying cost, but very little. At present the levels are all poor; the lode in the west shaft has underlayed faster than the shaft, and we have not seen it for several fathoms. The ground lately in the shaft has been cleaner killas, and if any alteration, better ground. It is now 9 fathoms under the 55-fathom level, and we are driving to cut the lode. The ground in the cross-cut is harder than when you were on the spot. The water is sinking in old Binner; it is about 7 fathoms under the adit in the western part, and deeper in the eastern part; we do not account for this. Wheal St. Aubyn combined poor. Wheal Abraham looks promising, and Creuver about paying cost. Dolcoath is better than when you left us, or when I was in London. The last sale was only about 800 tons. The next sale on Thursday is upwards of 1100 tons, and we expect a little better standard.

"I wish you could discover who that old gent is that wanted a large slice in Dolcoath, that I might get at him through some unknown channel, for I want money sadly.

"Cook's Kitchen continues poor, Tin Croft ditto; Wheal Fanny not rich. We had a pretty little fight last account there with T. Kevill and W. Reynolds, Esquires: black eyes and bloody noses the worst effects. T. Kevill's face was much disfigured, and he might have found a new road out of his coat.

"At a meeting of Condurrow adventurers yesterday, twenty-four of them agreed to have one of our engines, cylinder 12 inches in diameter and 6-feet stroke, provided the Foxes do not object to it. When the order is given I shall write to Mr. Hazeldine, provided I do not hear from you that it is better to send the order to any other place.

"If you have occasion to write Mr. Hazeldine, I wish you would press him to hasten the engines for Wheal Goshen, &c.

"I am served with a Vice-warden's petition by Mr. Harris

for not working the Weith mine in a more effectual manner, and he prays the Vice-warden to make the sett void. The trial will come on some time the beginning of July, and by that time I suppose we shall have two fire-engines working thereon.

"Had Mr. Harvey done as he was desired we should have had one working there at this time, but he has but now begun to do anything to it. We have the cylinder and ends home from Polgooth, and my cousin Simon Vivian is making the tubes. We have the other cylinder from Wheal Treasury, and I have ordered Horton to cast a cock for it the same as that at Dolcoath. We have cut the south lode at the adit level about 50 or 60 fathoms east of the engine, and have driven about 20 fathoms on it. It turns out about half a ton per fathom at 20*l.* a ton. The ground at 40*s.* per fathom; this all in a hole, and is better going down. The back is sett to four men at 3*s.* 11*d.*; their time is out this week, and I suppose they must have 5*s.* next. This may turn out a few thousands, and I think too promising a thing to give up to Mr. Harris.

"I am happy to inform you that all our friends are in good health, and beg my most respectful compliments to Mrs. Rogers and adopted son; and am,

"Dear Sir,

"Yours very sincerely,

"ANDW. VIVIAN.

"The promised news respecting the engine business I am very anxious to have, as it will I hope make me *proud*, as proud I shall be when I am able to pay everyone their demands, and have sufficient to carry on a little business to maintain my family and self without the assistance of others. May you succeed in your undertaking and also be independent, is the sincere wish of your friend. John Finnis and others are anxious to know when they will be wanted.   "A. V."

The explosion at Greenwich in 1803 was made much of, though the fault was clearly not in the boiler. Three years afterwards, in 1806, a steam-cylinder burst in Wales, therefore Mr. Williams, a large shareholder in Dolcoath, objected to the use of high-pressure expansive steam in their large Watt pumping engine, and desired their engineer, Mr. Sims, to make a competitive trial after his own fashion.   At Condurrow Mine one of Trevithick's engines was to be ordered if the Foxes and Williamses did not object; and so it was that Trevithick's high-pressure steam-boiler was not ordered, and the Watt vacuum engine was for a longer time to receive no increase of power.

"Some of Captain Dick's early boilers had flattish or oval fire-tubes.   In 1820 I repaired an old one in Wheal Clowance Mine in Gwinear.   The flat top had come down a little; we put in a line of bolts, fastening the top of the tube to the outer casing.

"About 1818 I saw in Carsize Mine in Gwinear a pumping engine that Captain Dick had put up.   The boiler was a cylinder of cast iron, with a wrought-iron tube going through its length in which the fire was placed.   The steam-cylinder was vertical, fixed in the boiler.   She had an air-pump and worked with a four-way cock.   The steam was about 100 lbs. to the inch."[1]

"About 1820 I removed one of Captain Trevithick's early high-pressure whim-engines from Creuver and Wheal Abraham, and put it as a pumping engine in Wheal Kitty, where it continued at work for about fifteen years.   The boiler was of cast iron, in two lengths bolted together, about 6 feet in diameter and 10 feet long.   At one end a piece was bolted, into which the cylinder was fixed, so that it had the steam and water around it.   There was an internal wrought-iron tube that turned back again to the fire-door end, where the wrought-iron chimney was fixed; the fire-grate end of the tube was about 2 feet 6 inches in diameter, and tapered down to about 1 foot 6 inches at the chimney end.   It was a puffer, working 60 lbs. of steam

---

[1] Banfield's recollections in 1869.

to the inch; it worked very well. There were several others in the county at that time something like it. It was made at the Neath Abbey Works in Wales."[1]

These boilers were of the kind first tried in Cornwall about 1800. The oval tube in the Kensington model of 1798 continued in use in Cornwall for many years. The cast-iron outer casing was soon abandoned, though one of them in Wales remained in work fifty years, using steam of 60 lbs. to 100 lbs. to the inch.

"My dear Jane,   "Hayle Foundry, *August 26th*, 1810.

"I saw Captain Andrew Vivian on Wednesday, who told me that he had been offered 150*l.* a year to inspect all the engines in the county, and report what duty they were doing, in order to stimulate the engineers. He declined accepting it, having too much to do already; and he thought it would be worth Trevithick's notice, as it would not take him more than a day or two in a month.

"I remain, my dear Jane,

"Yours sincerely,

"H. Harvey.

"I wrote this letter on Sunday, with an intention of sending it then, but thought it best to wait until this day, in hopes of hearing the determination of Government in your favour; but your letter has arrived without the desired information. All that I can now say is, to desire that Trevithick will make up his mind to return to Cornwall immediately.   "H. H."

The application to the Government for remuneration for benefits conferred on the public was unsuccessful. The office of registrar of Cornish engines was unsuitable; fortunately for mining interests, illness obliged Trevithick to revisit his native county, for by the increased power and economy of his engines Dolcoath Mine, so

---

[1] Recollections of Captain G. Eustace, engineer, residing at Hayle, 1868.

frequently mentioned, and so important in olden time, now returns 70,000*l.* worth of tin yearly.

Trevithick's first act on returning to Cornwall in 1810 was the erection of the high-pressure boilers and pole vacuum engine at Wheal Prosper; at the same time renewing his proposals to Dolcoath to use his improved boilers, which had been broken off in 1806, and to apply high-pressure steam to their low-pressure Watt engine, with the same safety and profit as in Wheal Prosper; the evidence was undeniable, so his plans were agreed to, and in the early part of 1811 the high-pressure boilers, called the Trevithick or Cornish boilers, were constructed in the Dolcoath Mine under his directions.

Old John Bryant, who worked the Dolcoath large engines both before and after the introduction of higher pressure steam, including the Carloose or Bullan Garden 45-inch cylinder engine, Wheal Gons 63-inch cylinder single engine, and the Watt 63-inch cylinder double, with the bee-but boiler, such as Trevithick, sen., used in 1775,[1] followed by the Watt waggon boiler, and afterwards by the globular boiler of Trevithick, jun., in 1799,[2] and still later also with the cylindrical boiler of 1811, gave the following statement, when seventy-four years old, to the writer :—

"In the old bee-but and the waggon boiler the steam pressure in the boiler was not much; we did not trouble about it so long as the engines kept going: when the steam was too high it blew off through the feed-cistern. When Captain Trevithick tried his high steam in Dolcoath we hoisted up the feed-cistern as high as we could; when the steam got up, it blew the water out of the cistern. Captain Dick holloed out, 'Why don't you trig down the clack?'

"The cylindrical boilers when they were first put in leaked

---

[1] See vol. i., p. 25.   [2] See vol ii., p. 119.

very much; we could hardly keep up the fire sometimes. I reckon the steam was 30 or 40 lbs. to the inch. Captain Dick's boilers made him lots of enemies. I heard say in one mine where he was trying his boilers against Boulton and Watt's waggon, a lot of gunpowder was put into the heap of coal." [1]

The waggon or hearse Watt boiler was attached to his 63-inch cylinder double, and the old man recollected having raised the water cistern, when Trevithick's globe boiler gave an increased pressure in 1799, ten or twelve years before the cylindrical boilers were made in Dolcoath.

"Some time after Captain Dick's globe boiler and steam-whims had been at work in Dolcoath, a letter came down from London, saying that he would save the mine 100*l.* a month if they would put in one of his new plan boilers.

"They were put in hand in the mine, and I worked about them; they were wrought-iron cylindrical boilers, about 20 feet long, and 5 or 6 feet in diameter; the fire-tube was about 3 feet in diameter; the fire returned around the outside in brick flues. Three boilers were put in side by side.

"When Captain Dick first tried them, he said to the men, Now mind, the fire-bars must never have more than six inches of coal on them; give a shovel or two to one boiler, and then to another. When Captain Dick's back was turned, the men said they wasn't going to do anything of the sort, there would never be no rest for them. They used to say that the boilers saved more than 170*l.* the first month." [2]

Clark, when a boy, in 1799, helped to construct Trevithick's globular boiler in Dolcoath, and recollected the events of the few following years, during the contests with the whim-engines about 1806, and the introduction of the large cylindrical wrought-iron boilers for the pumping engines in 1811, and the struggle pre-

---

[1] Old John Bryant's statement in 1858.
[2] Clark's recollections in 1869, when he was eighty-three years old, and resided at Redruth.

ceding the downfall of the Watt low-pressure steam
vacuum engine, to make room for the high-pressure
expansive steam-engine, with or without vacuum.

"About 1812 Captain Trevithick threw out the Boulton and
Watt waggon boilers at Dolcoath and put in his own, known as
Trevithick's boiler. They were about 30 feet long, 6 feet in
diameter, with a tube about 3 feet 6 inches in diameter going
through its length. There was a space of about 6 inches
between the bottom of the tube and the outer casing. Many
persons opposed the new plans. The Boulton and Watt low-
pressure engine did not work well with the high steam, and the
water rose in the mine workings. Captain Trevithick, seeing
that he was being swamped, received permission from the mine
managers to dismiss the old engine hands and employ his own
staff. Captain Jacob Thomas was the man chosen to put things
right. He never left the mine until the engine worked better
than ever before, and forked the water to the bottom of the
mine. Before that time the average duty in the county by
the Boulton and Watt engines was seventeen or eighteen millions, and in two or three years, with Trevithick's boilers and
improvements in the engines, the duty rose to forty millions.
About 1826 he (Captain Vivian) was manager of Wheal Towan;
their engines were considered the best in the county, doing
eighty-seven millions; they had Trevithick's boilers, working
with high-pressure steam and expansive gear; few if any of
Boulton and Watt's boilers could then be found in the county.
Sir John Rennie and other scientific men, who doubted the
reports of the duty, came and made their own trials with the
engines, and were satisfied that the duty was correctly reported.

"About that time a Mr. Neville requested him to report on
the engines at his colliery at Llanelthy; one was an atmospheric of Newcomen's, doing six millions; and four or five of
Boulton and Watt's patent engines averaged fourteen millions."[1]

When at last the cylindrical high-pressure boiler
was admitted, and men had been taught to fire them,

---

[1] Captain Nicholas Vivian was a schoolfellow and intimate friend of Trevithick's; he resided at Camborne in 1858, when he gave his recollections.

many persons still liked the old plans, and among them the easy-going low-pressure enginemen. The consequence was that the Watt engines under their management refused the early doses of Trevithick's high steam, not easily digesting it, and their obstinacy nearly swamped Trevithick and his plans.

" When a little boy, about 1812, I frequently carried my father's dinner from Penponds to Dolcoath Mine. One day, not finding him in the engine-house, I sought him in the account-house, but not knowing him in a miner's working dress, refused to give him his dinner. William West then worked with him. I heard there was difficulty in making the new boilers and the old engine work well ; engineers from other mines looked on from a distance, not liking the risk of explosion. People seemed to be against the new plans; some labourers worked with them."

This narration—sixty years after the events—from Mr. Richard Trevithick, the eldest son of the engineer, shows that William West helped in applying high-pressure steam to the Watt low-pressure engine, and that but few sympathized with the innovators on old customs; but among them was Captain Jacob Thomas, who successfully fed the old engine with strong steam.

At that time the Watt engines in Cornwall had been doing seventeen or eighteen millions; Trevithick's new boilers increased their duty to forty millions.

" William Pooly [1] was working in Dolcoath before Captain Trevithick's new boilers were put in, and helped to put them in.

" The Shammal 45-inch engine was an open-top cylinder, with a chain to the segment-head wooden beam. So was the 63-inch cylinder Stray Park engine, then called Wheal Gons [2]

---

[1] William Pooly worked the Dolcoath 76-inch engine in 1869; his recollections were given in the old engine-house, on the spot once occu-pied by Watt and his 63-inch great double engine.

[2] Smiles speaks of this as Bonze's.

in Dolcoath sett, and the Boulton and Watt 63-inch cylinder double-acting.

"There used to be great talking about different boilers; a boiler of Captain Trevithick's worked with higher steam than the others. Just before Captain Dick came back to the mine a Boulton and Watt hearse boiler had been repaired with a new bottom; it was never used. I and William Causan took a job to cut up the boiler at 1s. 6d. the hundredweight; it weighed 17 tons. Jeffrie and Gribble were the mine engineers; Glanville used to be considered Captain Dick's man in the mine. You could stand upright on the fire-bars in the middle hollow of the hearse boiler, and so you could in the outside brick flues; the middle hollow was like a horse-shoe. When Captain Dick put in his cylindrical boilers he altered the 63-inch single; there was hardly anything of her left but the main wall, with the wood bob and a chain to the piston-rod, and also to the pump-rods. There was an air-pump, and I think a second-hand cylinder was brought, but it was a 63-inch; the old Shammal engine had been altered, too.

"The new boiler put in was about 8 feet in diameter and from 30 to 40 feet long, two round tubes went through it; the fire-place in one end of one tube and in the other end of the other tube; after going through the tubes the draught went into the brick flues under the bottom and sides. When the new engine was put in, Gribble said, 'Why, these little things will never get steam enough;' everybody said so.

"In the Boulton and Watt engines we didn't trouble about feed-pumps and gauge-cocks.

"A wire came through a stuffing box in the top of the boiler; a biggish stone in the boiler was fastened to one end of the wire, the other end was fastened to a weighted lever near the water cistern, just above the boiler; when the water got low the stone opened the valve in the water cistern. That was when they were putting in Captain Dick's new cylindrical boilers to the old 63-inch engine. She did so much more work, with less coal, that in a year or so they agreed to throw out Boulton and Watt's engine, and to put in a stronger one that could stand Captain Dick's high steam. Jeffrie and Gribble were the mine engineers that put her up. The 76-inch cylinder came from Wales. The big beam was cast at Perran Foundry in 1815; you can see the name and date upon it now. The boiler and the gear-work were made in the mine. The exhaust-valve is exactly as when it was put in, worked by a rack-and-tooth segment. The equilibrium valve is unchanged, except that the rack is taken out and a link put in.

"The steam-valve was taken out soon after she went to work, and the present double-beat valve was put in; it is the first of the kind I ever saw. Some were made before that time with a small valve on the top of the big one, that opened first, to ease the pressure.

"John West[1] fitted up the valve-gear in the mine with the expansive tappets, the same as when she stopped a month or two ago, and the same as the present new one has.

"Captain Dick's cutting off his strong steam at an early part of the stroke, used to make the steam-valve strike very hard; so the new plan valve, with a double beat, was put in; that must have been about 1816 or 1817; and the valve and expansive horn for working were just exactly like what they have put into the present new engine in 1869. She was the engine that showed them how to fork the water, and burn only half the coal.

"I worked in this mine the old atmospheric engines, and then Boulton and Watt; and then Trevithick's boilers in Boulton and Watt; and then Trevithick's boilers and engine; and now I come every day to the new engine, though I can't do much. They give me 35s. a month; and my name is William Pooly, Dolcoath, 1869."

Three years ago (in 1869), when the writer entered the old engine-house in which Watt's 63-inch cylinder double had been erected in 1780, adjoining the old walls that then enclosed that early Newcomen 45-inch cylinder Carloose engine, re-erected by Trevithick, sen., in 1775 in Bullan Garden portion of Dolcoath,

---

[1] Three Wests, all skilful mechanical engineers, were employed at that time in Dolcoath, all of them known to the writer, who thinks the double-beat valve was the handiwork of John West, not related to Trevithick's partner.

an old man sat near a small window in a recess in
the thick wall of the engine-house, within reach of
the gear-handles of the Jeffrie and Gribble 76-inch
cylinder engine that Trevithick, jun., had erected in
1816 on the foundations of the removed Watt engine;
he held in one hand a portion of slate from the
roof, and in the other an old pocket-knife, one-
half of the blade of which had been broken off, leaving
a jagged fracture, with which he made the figures of
some calculation on the rude slate; on his nose rested
the brass frame of a pair of very ancient spectacles, with
horn glasses. He answered the writer's question by,
" Yes, I am William Pooly; I worked this engine, and
the other engines before it—the great double and the
little Shammal working out of the same shaft; and I am
seventy-four years of age. The 63 single worked upon a
shaft up there; she was called Wheal Gons." That old
man, still living, had worked in Dolcoath Mine one of the
first steam-engines of Newcomen; the 45-inch, modified
by Trevithick, sen.; then the 63-inch double of Watt;
and, finally, the high-pressure engines of Trevithick,
jun.; he saw the open-top cylinders, atmospheric of
Newcomen, in the Shammal 45-inch and Wheal Gons
63-inch, with their wooden beams with segment-headed
ends, moving in rivalry with the Watt 63-inch double,
with cylinder-cover and parallel motion; he saw the two
former engines, as altered by Trevithick, jun., using the
higher steam from the globular boiler on which Henry
Clark worked in 1799, when " there used to be great
talking about different boilers, and a boiler of Captain
Trevithick's worked with higher steam than the others;
and the waggon boiler of Watt, that had just been repaired, was discarded and cut up;" thus described by
Trevithick, " the fire-place is 22 feet from fire-door to

fire-door, 9 feet wide, and 7 feet thick in fire,"[1] which he proposed to replace in 1806 by a cylindrical boiler to give steam of 25 lbs. on the inch.

Pooly also saw the finishing stroke in 1811, when the boilers still known as the Trevithick or Cornish boilers, gave steam to the three engines ; after a twelve years' fight between low and high pressure, commencing with Trevithick's globular boiler and internal tube, in Dolcoath, in the year 1799, from which time it gained step by step, though in comparatively small engines, up to 1811, when the cylindrical boilers took the place of the condemned hearse and globular boilers, and gave really strong and expansive steam to the three Dolcoath pumping engines that from time immemorial had been rivals, causing all three of them to lift an increased quantity of water, and at the same time to save one-half in the cost of coal; this continued for four or five years, when in 1816 the 63-inch double and the 45-inch, being the youngest and the oldest of the three, were removed, that a new 76-inch cylinder, better adapted to Trevithick's expansive steam might more cheaply perform their joint work. Prior to this change the three engines were known by the names Shammal 45-inch, formerly Bullan Garden,[2] but before that as Carloose, of the period and form of the Pool engine;[3] Stray Park 63 single, formerly Wheal Gons,[4] dated from 1770 to 1777; and the 63-inch double of Watt in 1780.

" SIR,   " DOLCOATH MINE, *March 29th*, 1858.

"I have obtained the following information respecting the building of the first cylindrical boilers, as ordered by your late father for Dolcoath; and some information of the results

---

[1] See Trevithick's letter, February 18th, 1806, vol. ii., p. 143.
[2] See vol. i., p. 25.   [3] See vol i., p. 5.
[4] Query Bonze, spoken of by Smiles.

as to the coals consumed, compared with the consumption by the boilers previously in use here.

" George Row, now about seventy-two years old, and working at Camborne Vean Mine, says he assisted to build the two first cylindrical boilers with internal tubes used in Cornwall. They were built in Dolcoath Mine in the year 1811; they were 18 feet long, 5 feet diameter, having an oval tube 3 feet 4 inches in the largest diameter at the fire end; the other or chimney end of the tube was somewhat smaller. They were found too small for the work to be done, and another boiler was built immediately, 22 feet long, 6 feet 2 inches diameter, and he believed a 4-feet tube.

" John Bryant, now seventy-four years old, works a steam-engine at West Wheal Francis. He worked at Dolcoath the 63-inch cylinder double-acting engine, upon Boulton and Watt's plan. When he first worked her she had the old bee-but boiler, 24 feet in diameter. They were taken out for the Boulton and Watt waggon boiler, 22 feet long and 8 feet wide, with two fire-doors opposite one another.

" Then the Boulton and Watt waggon was taken out for Captain Trevithick's boilers, which he worked for several years. Two boilers were put in, each 18 feet long, 5 feet diameter, with an internal oval tube, he thinks, 3 feet by 2 feet 6 inches. Shortly after, another boiler of similar form was added, 22 feet long, 6 feet diameter, 4-feet tube.

" He cannot say what the saving of coal was, but remembers that the duty performed by the engine with the waggon boiler was thirteen to fourteen millions. Mr. William West came to the mine as an engineer, and by paying great attention increased the duty of the Boulton and Watt engine and boiler to about fifteen millions. He does not recollect the duty the engine performed with the cylindrical boilers.

" Mr. Thomas Lean, of Praze, the present reporter of mine engines in the western part of Cornwall, in answer to a note I sent to him, says he has no account of any report of Dolcoath engines for the *year* 1812, but during the month of April in that year the engines did $21\frac{1}{2}$ millions. During the whole of 1813 that engine was reported to average a duty of twenty-one

M 2

millions. The whole of the above are at per bushel of 93 lbs., and the whole of the accounts furnished by Mr. Lean are for Trevithick's cylindrical boilers.

"From the Dolcoath Mine books I find the following: Paid for coals for the whole mine during the year 1811, 1150*l.* 15*s.* 10*d.*, or per month, 931*l.* 14*s.* 7*d.* During the first three months of 1812 the coal averaged 1000*l.* per month. In May of this year, 1812, Captain Trevithick is entered on the books as paid 40*l.* on account of boilers; and in August of the same year, for erecting three boilers, 105*l.* I think the three boilers were at work in April, 1812, the month Mr. Lean gives as the first reported. From April, 1812, to December, during nine months, the cost of coals was 5512*l.* 6*s.*, averaging 612*l.* 9*s.* 6*d.* per month. During the next year, 1813, the cost for coal was 7019*l.* 17*s.* 5*d.*, or an average per month of 590*l.* 16*s.* 5*d.* I cannot find the price paid per ton for the coal in these years, but the average price during 1808 and 1815 was much alike, making it probable that the price per ton during 1811, 1812, and 1813, was nearly the same; and that the saving of the above 300*l.* per month in Dolcoath was wholly on account of the saving effected by Trevithick's cylindrical boilers.

"The testimony of John Bryant, that the duty with the waggon boiler was say fourteen millions, and that of Mr. Lean, giving twenty-one millions with the new Trevithick boiler, bear much the same proportion as the charges for coals in the respective periods above given.

"In the year 1816 a new 76-inch single engine was erected in the place of the old Boulton and Watt 63-inch double engine with Trevithick's cylindrical boilers. The average duty performed during the year 1817 was 43¾ millions. This same engine is still at work, and her regular duty is from thirty-six to thirty-eight millions.

"I am, Sir,

"Your most obedient servant,

"CHARLES THOMAS.

"FRANCIS TREVITHICK, Esq."

Captain Charles Thomas, who was one of the most experienced of Cornish miners, for many years the manager of Dolcoath, and in youth the acquaintance of Trevithick, states that the new high-pressure boilers were made in the mine in 1811, and gave their first supplies of strong steam to the three large pumping engines in April, 1812, with such good effect that the increasing water which had threatened to drown the mine was speedily removed, and that with a saving of nearly one-half of the coal before consumed. Prior to their use Dolcoath Mine paid 1000*l.* monthly for coal; but for the latter nine months of the year, in consequence of the new boilers, the cost was reduced to 612*l.* a month. This saving in the pumping cost of one mine crowned with success the high-pressure steam engineer, who had been steadily gaining ground during his fight of twelve or fourteen years on the battle-ground chosen by Watt thirty-three years before.

The low price of tin and copper, which caused so many engines to cease working about the close of the last century, had changed for the better, and the present century opened with an increasing demand for steam power. Trevithick's high-pressure portable engines had worked satisfactorily for several years; and as a means of making public the relative duty performed by Cornish pumping engines, and of solving conflicting statements on the rival systems of low and high pressure steam, it was determined that an intelligent person should examine and give printed monthly reports of the amount of duty done by the different engines, and in 1810 Captain Andrew Vivian was requested to take this work of engine reporter in hand; on his refusal it was offered to Trevithick. In August, 1811, Mr. Lean commenced such monthly reports, showing that the duty of twelve pumping engines at the end of that year averaged seventeen millions, exactly the duty done by the

Boulton and Watt engines thirteen years before, as reported by Davies Gilbert and Captain Jenkin in 1798, proving the small inherent vitality of the Watt engine.

In 1814 the Dolcoath pumping engines, with Trevithick's cylindrical boiler and high steam expansion, are thus reported :—"The Boulton and Watt, Dolcoath great double engine, 63-inch cylinder, did a duty of 21½ millions; the Shammal 45-inch cylinder, single engine, did 26¾ millions; and the 63-inch single, Stray Park engine, 32 millions." Shammal engine, nearly 100 years old, beat the Watt engine of more than half a century later; and so did Stray Park 63-inch, which Watt had laughed at when he first tried his hand as an engineer in Cornwall in 1777.[1]

The marked change in these three engines, while for two or three years under Trevithick's guidance, becoming more powerful and economical, raised the usual swarm of detractors, and in 1815 a special trial was made, which lasted for two days, to test the reported increased duty by the cylindrical boilers and expansive working.

The unbelievers were then convinced, and agreed to throw out the Boulton and Watt great double engine 63-inch cylinder, together with its neighbour, the worn-out old 45-inch, and put in their stead one engine with a cylinder of 76 inches in diameter, with expansive valve and gear, and parts strong enough and suitable to the high-pressure steam, on Trevithick's promise that it should do more than the combined work of the other two with one-half the coal.

In 1816 this new engine commenced work, and did forty millions of duty, increasing it during the next

---

[1] See vol. i., p. 30; vol. ii., p. 115.

THE WATT AND THE TREVITHICK ENGINES. 167

two or three years to forty-eight millions, being three
times the duty performed by the Watt 63-inch double
engine before it was supplied with steam from Trevithick's boilers, and twice as much as it performed when
so supplied. Lean says, "This was the first instance of
such duty having been performed by an engine of that
simple construction." The other mines followed Trevithick's advice, but never paid him a penny. On this
Lean again says, "The engines at work in the county in
1835 would have consumed 80,000*l.* worth of coal over
and above their actual consumption yearly, but for the
improvements that had been made since 1814."

Trevithick's engines were very durable, as well as
cheap in first cost and in working expense. This
famous Dolcoath 76-inch engine remained in constant
work night and day for fifty-four years; after which
good service the steam-pipes, being thinned by rust,
were held together by bands and bolts; the steam-case
around the cylinder would no longer bear the pressure
of steam; the interior of the cylinder from wear was one
inch larger in diameter than when first put in, and had
to be held together by strap-bolts. The original boilers
were said to remain, only they had been repaired until
not an original plate remained; but there they were in
the old stoke-hole in 1869, when, from the fear of some
part of the engine breaking and causing accident, it
was removed.

In 1867 the writer was a member of the Dolcoath
Managing Committee, when it was determined that the
old engine of 1816 should be replaced by a new one.
The cylinder sides were reduced in thickness by half an
inch; the steam-pipes and nozzles were thinned by rust
and decay; the valves and gear-work remained in good
order. Captain Josiah Thomas, the present manager of

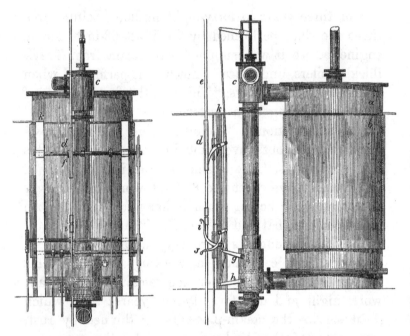

TREVITHICK'S DOLCOATH 76-INCH CYLINDER PUMPING ENGINE, ERECTED IN 1816, CEASED
WORKING 1869.

a, steam-cylinder, 76 inches in diameter, 9-feet stroke; *b*, steam-jacket; *c*, steam expansion-valve,
11 inches diameter, double beat; the u̟ per beat 11 inches diameter, the under beat 9½ inches, valve
8 inches long; *d*, expansive cam on plug-rod; *e*, plug-rod for moving the gear; *f*, expansive horn;
g, equilibrium valve, 13 inches in diameter, single beat moved by a tooth-rack and segment; *h*, exhaust-valve, 14½ inches in diameter, single beat moved by a lever and link; *i*, equilibrium-valve
handle; *j*, exhaust-valve handle; *k*, Y-posts for carrying the gear arbors; *l*, main beam in two plates
of cast iron; *m*, parallel motion; *n*, feed-pump rod; *o*, air-pump bucket-rod, the pump, 2 feet 9 inches
diameter; *p*, the main pump rods.

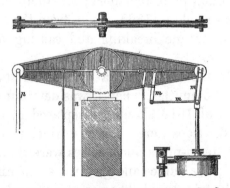

CYLINDER, MAIN BEAM, AND PUMP-ROD OF DOLCOATH 76-INCH CYLINDER ENGINE.

the mine, offered to sell this old engine at scrap price, that it might be stored in the Patent Museum at Kensington as a memento of the early high-pressure expansive steam pumping engine.

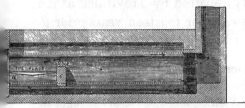 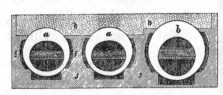

BOILERS ERECTED IN 1811 IN DOLCOATH, USED IN THE BOULTON AND WATT 63-INCH ENGINE, THEN IN THE NEW 76-INCH UNTIL 1869.

a a, two wrought-iron cylindrical boilers, 5 feet in diameter, 18 feet long, with internal fire-tube, oval, 3 feet 4 inches by 3 feet; *b*, a boiler, 6 feet 2 inches diameter, 22 feet long, cylindrical tube, 4 feet diameter in the fire-place, the remainder 3 feet; *c*, brick bridge; *d*, fire-bars; *e*, brick external flues under boiler; *f*, brick side-flues; *g*, ashes, or other non-conductor; steam 30 to 50 lbs. on the inch above the atmosphere.

The steam-cylinder of 1816 was cast in South Wales; the beam still working in the new engine of 1869 was cast in the foundry of the Williams' at Perran. John West replaced the original flat expansive steam-valve with a double-beat valve; the gear was principally made by him on the mine, and remained in good working to the last. This double-beat valve is the first the writer has met with; it is of the same form as the modern double-beat valve; an earlier plan was to have a small valve on the top of the main valve. The steam in ordinary working was shut off when the piston had moved from an eighth to a quarter of its stroke.

The Gons, or Stray Park 63-inch cylinder, survived its companions, the 63 double, and 45 single, for some ten or fifteen years, having beaten both of them in duty. A memorandum in Trevithick's handwriting shows that he in 1798, when designing his large globular boiler with internal flue at the reworking of Dolcoath, tested the relative duty of the Watt 63-inch double and the 63-inch single engine, then called Wheal Gons, the latter in its original form of open-top cylinder atmospheric; shortly after which it probably received a cover about the same time as the 45-inch, for both those engines were thoroughly repaired by Trevithick at the reworking of the mine, twelve or fourteen years prior to the use of the cylindrical boilers.

"At the time that Boulton and Watt made their trial of Seal-hole engine against Hornblower's engine at Tin Croft, the engines were put in the best order, and good coals brought in for the purpose, to work for twenty-four hours. The trial was attended by the principal mining agents; the result was about ten millions by each engine.

"At Dolcoath Mine an old atmospheric engine continued to work for several years by the side of one of Boulton and Watt's engines of the same size; the water lifted and coals consumed were carefully taken and made known to the public, showing that Boulton and Watt's engine performed, when compared with the old engine, as 16 to 10."[1]

Hornblower was an active engineer in Cornwall before Watt; the patent of the latter claiming the sole right of working an engine by steam in the cylinder,[2] drove the former to use two cylinders, in one of which the expansion was carried out, as a means not described in Watt's patent; a lawsuit was the consequence. The two engines when tried by Trevithick[3] performed an equal duty of ten millions. In 1798 he tested the Dolcoath atmospheric 63-inch single against Watt's great 63-inch double action. "The atmospheric performed ten millions," precisely the duty of the patent Watt and the patent Hornblower contests of six years before; but the Watt Dolcoath engine, then considered the best he had

---

[1] Memorandum in Trevithick's writing.   [2] See vol. i., p. 46.
[3] See vol. i., p. 57.

made, did sixteen millions. These trials in 1792 and 1798 enable us to compare the Newcomen, the Horn-blower, and the Watt engines; shortly after which Trevithick tried higher steam in one or more of those same engines from his globular boiler.[1]

"SIR CH. HAWKINS,   "CAMBORNE, *March 10th*, 1812.

    "Sir,—This day I shall attend the account at Wheal Prosper Mine, in Gwythian, to contract with the adventurers for erecting a steam-engine on my improved plan, for drawing the water 50 fathoms under the adit. I called on Wheal Liberty adventurers at St. Agnes last week, and found that several of them had given up their shares rather than put in a new engine, and the remainder of them very sick.

    "I told them that I would fork the water with the present engine, and draw instead of 40 gallons each stroke, 47 fathoms deep (which she did), 85 gallons per stroke, 65 fathoms deep, by altering the engine on the same principle as I have done with the Dolcoath great engine, and several more that are now altering. The expense of altering the engine, and forking the water to bottom, and proving the mine, will not exceed 1000*l.*

    "All the adventurers are very anxious to again resume their shares and make the trial, on condition that I will undertake the completion of the job at a certain sum, but not otherwise.

    "I am certain, from what Dolcoath engine is doing, that I can far exceed the power above stated, and perform the duty with one-half the coal the engine consumed before, and would not hesitate a moment to engage the job on the terms they propose, but I have not money sufficient to carry it into execution, as I must lay out a large sum in erecting the engine on the Gwythian Mine, and unless I can be assisted with 500*l.*, shall not be able to undertake the job.

    "If you think it worth your notice to encourage this undertaking by lending me the above sum for six months, I will pay you interest for it, and before drawing any part of it from you would get materials in the mine that should amount to above

---

[1] See vol. ii. p. 119.

that sum, and also give you an order on the adventurers to repay you the whole sum before receiving any part myself.

" As I have been a bankrupt, perhaps you may scruple on that account, but that business is finally settled, and I have my certificate; and indeed I never was in debt to any person; not one shilling of debt was proved against me under the commission, nothing more than the private debts of my swindling partner.

" At Wendron we are working an engine lately erected on a copper lode, which has a very promising appearance, and near this spot you have land at Besperson. where there is also a very kindly copper lode, which deserves trial; if you are inclined to grant a sett, I think I can find adventurers to join me to try the mine.

" I have lately read a letter from your hind, that the engine continues to mend; it far exceeds my expectation. I am now building a portable steam-whim, on the same plan, *to go itself* from shaft to shaft; the whole weight will be about 30 cwt., and the power equal to twenty-six horses in twenty-four hours.

" The only difference in this engine and yours will be the fire in the boiler, and without mason-work, on account of making it portable. I shall pass the rope from the fly-wheel round the cage of the horse-whim.

" If you should fall in with any West India planter that stands in want of an engine, he may see this one at work in a month, which will prove to him the advantage of a portable engine, to travel from one plantation to another. The price, completely finished and set to work, free of all expense, in London, 105*l*.

" I am, Sir,

"Your very humble servant,

"RD. TREVITHICK.

" N.B.—Captain John Stephens informed me, a few days since, that the lead mine at Newlyn was rich."

In Wheal Prosper Mine the first high-pressure expansive steam-condensing pole-engine had been worked, just before the date of the foregoing letter, and that

evidence of increased power and economy was immediately followed by the application of the same principles of high-pressure steam and very expansive working to the Watt low-pressure steam vacuum engines at
Wheal Alfred, Dolcoath, and other mines, with such
satisfactory results as to warrant his offering, on the
battle-ground of his first attack on the Watt low-pressure steam vacuum principle at Seal-hole in St. Agnes,
fourteen years before,[1] at his own pecuniary risk, to so
apply those principles in the Wheal Liberty low-pressure
steam-engine, which had failed to drain the mine, lifting only at the rate of 1880 gallons of water one
fathom high at each stroke ; that it should lift an
increased quantity of water, and that, too, from an increased depth, making the load equal to 5525 gallons,
and to perform such increase of work with one-half of
the quantity of coal before used ; in other words, he
was willing to engage to make the old low-pressure
steam-engine perform by its conversion into a high-
pressure steam-engine threefold its original work, and
also to increase its duty or economic value sixfold ; resting his argument on the similar changes, then to be
seen in operation at Wheal Alfred Mine, and especially
in the Watt 63-inch double-acting engine at Dolcoath,
whose history we have been tracing.   Well might Sir
Charles Hawkins hesitate to believe what the experience
of sixty years has barely sufficed to make plain to us.

"CAPTN. TREVITHICK,   "PENZANCE, *March 27th*, 1813.

    "Sir, — In consequence of the conversation that has
passed between you and West Wheal Tin Croft adventurers,
the said adventurers have resolved to put an engine on that
mine, agreeable to the proposals offered by you; that is, the
engine shall be capable of lifting a 5-inch bucket, 50 fathoms,

---

[1] See vol. i., p. 90.

4-feet stroke, 15 strokes per minute, or a duty equal thereto; for which they will pay you 50 guineas one month after the engine shall be at work, and 50 guineas more at four months after that, and 50 guineas more at four months from that time, making the full payment of 150 guineas in nine months from the time the engine shall set at work, the adventurers paying all expense, except the engine materials, which shall be delivered on the mine. But in case the engine not performing the above duty, the adventurers to be at liberty to return the same engine, and you to pay back all the money that you had received for the said engine.

"Signed by GABL. BLEWETT,
"in behalf of the Adventurers and Company."

Trevithick was willing to spend more than his last penny in establishing the superiority of his high-pressure steam expansive engines, but the selfishness of adventurers retarded their progress. The atmospheric, mentioned by Watt as working in Dolcoath in 1777,[1] did five or six millions of duty, yet in Trevithick's hands, about 1798 to 1800, when he erected his globular boiler with internal tube, one of them was tested with the 63-inch Watt low-pressure vacuum engine, when the latter did sixteen millions to ten millions by the atmospheric engine, being nearly double the duty it performed in its original form; and we shall still trace this same engine as Bonze or Gons until it increased to six times its first duty under the name of Stray Park 63-inch.

Trevithick having erected a high-pressure steam condensing whim-engine at Cook's Kitchen,[2] and in Dolcoath[2] a high-pressure puffer whim-engine, pleaded hard in 1806[3] to be allowed to supply the large pumping engines of Newcomen and Watt with higher pressure steam from

---

[1] See vol. i., pp. 30, 57 ; vol. ii., p. 115.   [2] See vol. i., p. 91.
[3] See vol. ii., p. 142.

his cylindrical boiler, which after years of consideration
Dolcoath, in 1811, agreed to. In 1813 he wrote:—
"That new engine you saw near the sea-side with me
is now lifting forty millions, one foot high, with one
bushel of coal, which is very nearly double the duty
that is done by any other engine in the county. A few
days since I altered a 64-inch cylinder engine at Wheal
Alfred to the same plan, and I think she will do equally
as much duty. I have a notice to attend a mine meeting
to erect a new engine, equal in power to a 63-inch
cylinder single."[1]

The beneficial results of those acts are too large to
be here entered into in detail. In round numbers, the
early pumping engines of Newcomen did five millions;[2]
Trevithick caused them to do ten millions of duty with
a bushel of coal. Watt, during thirty years of improvements, caused the duty to reach sixteen or twenty
millions in 1800. Trevithick, on the expiry of the
Watt patent, then came into play, and before he had
reigned half the time of Watt, again doubled the
duty of the steam-engine, as he states in 1813 "his
new engine was doing forty millions, being nearly
double the duty of any other engine in the county."
These statements by Trevithick agree very nearly with
the generally-received accounts of the progressive duty
of the large pumping steam-engine.

"In 1798 Davies Gilbert, Esq., and the late Captain Jenkin of
Treworgie, found the average of the Boulton and Watt engines in
Cornwall to be about seventeen millions. In August, 1811, the
eight engines reported averaged 15·7 millions. During the
year 1814 Dolcoath great engine, with a cylinder of 63 inches in
diameter, did twenty-one and a half millions nearly. Dolcoath

---

[1] See Trevithick's letter, January 26th, 1813, vol. ii., p. 55.
[2] See vol. i., p. 41.

Shammal engine, with a cylinder of 45 inches in diameter, did twenty-six and three-quarter millions. Dolcoath Stray Park engine, with a cylinder of 63 inches in diameter, did thirty-two millions.

" In 1815 a trial was made, to prove the correctness of the monthly reports. Stray Park engine at Dolcoath was chosen for the purpose, because its reported duty was such as led some persons to entertain doubts of its accuracy. The trial was continued for ten days, to the full satisfaction of all concerned.

" In 1816, Jeffrie and Gribble erected a new engine, 76-inch cylinder, single, at Dolcoath, which did forty millions. This was the first instance of such duty having been performed by an engine of that simple construction.

" In 1819, Dolcoath engine performed the best during this year, and at one time reached forty-eight millions.

" In 1820, Treskerby engine, to which Trevithick's high-pressure pole had been adapted, reached 40·3 millions.

" In 1816, Sims also erected an engine at Wheal Chance, to which he applied the pole adopted by Trevithick in his high-pressure engines. This engine attained to forty-five millions.

" In 1828 public attention had now been attracted to the improvements which Captain Grose had introduced into his engine at Wheal Towan. The duty of this engine, in the month of April this year, equalled eighty-seven millions.

" This again gave rise to suspicions of error in the returns. This engine was accordingly subjected to a trial (as Stray Park engine had been in 1815), which was superintended and conducted by many of the principal mine agents, engineers, and pitmen of other mines.

" The quantity of coal consumed in 1835, compared with the quantity that would have been consumed by the same engines in the same time, had they remained unimproved from the year 1814, shows that the saving to the county amounts to 100,000 tons of coal, or 80,000l. sterling per annum." [1]

Lean seems to have calculated on a bushel of coal as 94 lbs. In 1798, when Trevithick was about to give

---

[1] Lean's ' Steam-Engine in Cornwall.'

increased pressure of steam to the Cornish engines, his friend Davies Gilbert reported the average duty of the Watt engine in Cornwall to be seventeen millions.

In August, 1811, the reported duty averaged 15·7 millions. This was the month and year in which Trevithick, after twelve years of working evidences of the reasonableness of his promises of increased power and economy from using high-pressure steam, was allowed to erect his cylindrical boilers for the large pumping engines in Dolcoath Mine.

Has the reader realized that the 45-inch atmospheric Carloose engine, of nearly 100 years before,[1] had in 1775[2] become the Bullan Garden engine of Trevithick, sen., which was improved and re-erected by Trevithick, jun., in 1799,[3] when the name was again changed, this time to Shammal, because it was linked to another engine, no other than the Watt 63-inch double engine? This Shammal 45-inch took steam from the globular boiler, using a pole air-pump[4] and a Watt condenser, though retaining the beam with the arched head and chain connection; and again in 1811 took still more highly expansive steam from the cylindrical boilers with a new beam and parallel motion, enabling it in 1814 to beat its rival, the Watt Dolcoath great double engine.[5] The old 63-inch Gons, under the name of Dolcoath Stray Park engine, with Trevithick's improvements, did sixty-seven per cent. more work than the Watt 63-inch with an equal quantity of coal.

This startling fact was disbelieved by the advocates of low-pressure steam, and as the visible change in the Dolcoath engine from Newcomen to Watt, and from Watt to Trevithick, had been gradual and not very

---

[1] See vol. i., p. 21.   [2] See vol. i., p. 25.   [3] See vol. ii., p. 120.
[4] See vol. ii., p. 122.   [5] See Lean's report, vol. ii., p. 175.

striking, and the public were careless of principles, the one most puffed was most thought of; but the money saved was tangible, and in 1815 a special trial was made, which lasted two days, to discover if it was really true that Trevithick's appliances could so increase the duty of the engine. The 63-inch cylinder, then called Stray Park engine, was selected; the result proved that the large saving reported from Trevithick's boilers and expansive working during the last three or four years, was an incontrovertible fact.

The high-pressure steam was also given to the defeated Watt 63-inch double engine; yet this newest of the three engines was the first to be condemned, and her place was taken in 1816 by an engine of 76 inches in diameter, which Trevithick promised should, with his high steam and new expansive gear, do the work of the Watt 63-inch and the old 45-inch put together; which was more than fulfilled by its doing forty millions, and, as Lean says, " was the first instance of such duty having been performed by an engine of that simple construction."

In 1819 the new 76-inch engine which had been erected by the mine engineers, Jeffrie[1] and Gribble, who had long been employed by Trevithick in Dolcoath, was the best in the county, doing forty-eight millions, nearly three times the duty as given by Mr. Gilbert for the Watt engine in 1798. In 1827 Trevithick's pupil, Captain Samuel Grose, erected his Wheal Towan engine, which performed a duty of eighty-seven millions, some of the working drawings of which were made by the writer. In 1835 the principle laid down by Trevithick had become so general in the county as to cause a saving

---

[1] See vol. i., p. 106.

to the Cornish mines, in coal alone, of 80,000*l*. yearly. In addition to this, the increased power of the engine lessened the first cost by at least one-half.

The national importance of such weighty facts calls for further corroborative proof, for we can scarcely believe that two atmospheric low-pressure steam-engines, made before the time of Watt, could be altered so as to perform more work, and at a less cost than the Watt engine, by an ingenious supply of higher steam pressure from Trevithick's boilers, together with the Watt air-pump and condenser. The following words from Watt are descriptive of his practice, though contrary to his patent claim :—

"At a very early period, while experimenting at Kinneil, he had formed the idea of working steam expansively, and altered his model from time to time with that object. Boulton had taken up and continued the experiments at Soho, believing the principle to be sound, and that great economy would attend its adoption.

"The early engines were accordingly made so that the steam might be cut off before the piston had made its full stroke, and expand within the cylinder, the heat outside it being maintained by the expedient of the steam-case. But it was shortly found that this method of working was beyond the capacity of the average enginemen of that day, and it was consequently given up for a time.

"'We used to send out,' said Watt to Robert Hart, 'a cylinder of double the size wanted, and cut off the steam at half-stroke.'

"This was a great saving of steam, so long as the valves remained as at first; but when our men left her to the charge of the person who was to keep her, he began to make, or try to make, improvements, often by giving more steam. The engine did more work while the steam lasted, but the boiler could not keep up the demand. Then complaints came of want of steam, and we had to send a man down to see what was wrong.

"This was so expensive, that we resolved to give up the

expansion of the steam until we could get men than could work
it, as a few tons of coal per year was less expensive than having
the work stopped. In some of the mines a few hours' stoppage
was a serious matter, as it would cost the proprietor as much as
70*l.* per hour." [1]

Pole expresses the same view, intimating that Watt
only used steam of 1 or 2 lbs. pressure to the inch.

" In Watt's engine, as is well known, the pressure of steam in
the boiler very little exceeded the pressure of the atmosphere.
He recommended that when the engine was underloaded, this
excess should be equal to about 1 inch of mercury; and when
full loaded, ought not to exceed 2 inches; adding, ' It is never
advisable to work with a strong steam when it can be avoided,
as it increases the leakages of the boiler and joints of the steam-
case, and answers no good end.' [2]

" Mr. Watt's engines with such boilers" (which will not
retain steam of more than $3\frac{1}{8}$ lbs. per square inch above the
atmosphere) " cannot be made to exert a competent power to
drain deep mines, unless the supply of steam to the cylinder
is continued until the piston has run through more than half its
course.[3]

" In 1801-2 Captain Trevithick erected a high-pressure
engine of small size at Marazion, which was worked by steam
of at least 30 lbs. on the square inch above atmospheric pressure.
In 1804, as Mr. Farey admits,[4] the same gentleman introduced
his celebrated and valuable wrought-iron cylindrical boilers,[5] now
universally used in this county.

" To these, everyone at all acquainted with the Cornish
improvements ascribes a great part of the saving we have
obtained. This will further appear from an extract from a
valuable work edited by John Taylor, Esq., F.R.S.[6]

" The monthly consumption of coal in Dolcoath Mine was,

---

[1] Smiles' ' Lives of Boulton and Watt,' p. 228.
[2] Appendix A to Tredgold, 'Pole on Cornish Engines,' p. 49.
[3] 'Phil. Mag. and Annals,' N.S.,
vol. viii., p. 309, by W. J. Henwood.
[4] Ibid., p. 313.
[5] Ibid., vol. i., p. 127.
[6] ' Records of Mining,' p. 163.

in 1811, 6912 bushels; in 1812, 4752 bushels.[1] The alteration
in the boilers was the introduction of Captain Trevithick's cylindrical boilers in the place of the common waggon boilers, which
had until then been there in use.

"Mr. Woolf, as Mr. Farey states, came to reside in Cornwall
about the year 1813, and his 'first engines for pumping water
from mines were set up by him in 1814.'"[2]

The foregoing was read at the Philosophical Society
in 1831, to refute erroneous statements on the Watt
and Trevithick engines.. My friend Mr. Henwood had
at that time made official experiments in conjunction
with Mr. John Rennie on the detail, working, and duty
of high-pressure steam Cornish engines, the Watt low-
pressure steam principle having been wholly given
up. Rees's 'Cyclopædia'[3] also bears the following
similar testimony to date of the increased duty :—

"Trevithick's high-pressure engine was erected in Wales in
1804 to ascertain its powers to raise water. The duty was
seventeen millions and a half pounds raised one foot high for
each bushel of coals.

"The high-pressure steam-engines require a greater quantity
of coals, in proportion to the force exerted, than the engine of
Mr. Watt, and consequently are not worked with advantage in
a situation where coals are dear.

"From the reports of the engines now working in the mines
of Cornwall, which, with the exception of a few of Woolf's
engines, are all on Mr. Watt's principle, and most of them constructed by Messrs. Boulton and Watt, taking the average of
nine engines—bad, good, and indifferent together—they were
found in August, 1811, to raise only thirteen millions and a half.
But when it was known by the engine keepers that their engines
were under examination, they took so much pains to improve the

---

[1] Alteration in the boilers that
year.
[2] 'Phil. Mag. and Annals,' vol. x.,
p. 97, "Notes on Some Recent Improvements of the Steam-Engine in
Cornwall," by W. Jory Henwood,
F.G.S.
[3] See Rees ' On the Steam-Engine,'
published 1819.

effects, that by gradual increase the engines in 1815 lifted twenty-one millions and a half, taking the average of thirty-three engines. In 1816, Stray Park, a 63-inch cylinder, 7 feet 9 inches stroke, single-acting, being one of the three engines on the vast Dolcoath Mine; its performance in four different months was thirty-one, thirty-one and a quarter, twenty-eight, and twenty-eight and a half millions."

This statement reveals a source of error in estimating the relative values of the Watt and the Trevithick engine; that of the latter was the Welsh locomotive, compared in duty with the large Watt pumping engine, pointed out in Trevithick's letter[1] of that time, as an unfair comparison; the small high-pressure puffer, in 1804, is admitted to have done seventeen and a half millions of duty with a bushel of coal of 84 lbs., while in Rees' calculation of the engines, he gives Watt 94 lbs. of coal to a bushel; and having stated that the Watt pumping engines in Cornwall, in 1811, averaged but thirteen and a half millions of duty, draws the false conclusion that the high-pressure cannot compete with the low-pressure where coals are dear; yet he agrees with other writers that the great increase in the duty of the Cornish pumping engines commenced from 1811 (when Trevithick first gave them his high-pressure steam); and states that in 1816 the Stray Park 63-inch cylinder single-acting engine,[2] being one of three then working in Dolcoath, did thirty-one millions.

The 'Encyclopædia Britannica' on the question of duty states:[3]—

"The duty of the best of Smeaton's engines was, in 1772, 9,450,000 foot pounds per cwt. of coal. On the expiration of

---

[1] See vol. i., p. 166.   [2] See Watt's statement, vol. ii., p. 115.
[3] See "Steam-Engine," published 1860.

Watt's patent, about the year 1800, the highest duty of his engines amounted to twenty millions, or more than double the former duty, which may represent the economical value of the improvement effected by Watt under his various patents.

" The reported duty of Cornish pumping engines, by the consumption of 94 lbs. of coals, rose from an average of nineteen millions and a half, and a maximum of twenty-six millions in 1813, to an average of sixty millions and a maximum of ninety-six millions in 1843. It is necessary to bear in mind the distinction between the duty of a bushel of coal and 112 lbs."

Here, also, are the same general facts as to the duty of the Watt engine, and the marked and rapid increase of duty dating from Trevithick's Dolcoath engines in 1811; but the confusion and even contradictions in the statements prove how little the subject was understood.

" A rough draft, prepared by Mr. Edmonds on Trevithick's return from America, dated 1828, for an application to Parliament for remuneration to Trevithick, says, ' That this kingdom is indebted to your petitioner for some of the most important improvements that have been made in the steam-engine.

" ' That the duty performed by Messrs. Boulton and Watt's improved steam-engines in 1798, as appears by a statement made by Davies Gilbert, Esq., and other gentlemen associated for that purpose, averaged only fourteen millions and a half (pounds of water lifted 1 foot high by 1 bushel of coal), although a chosen engine of theirs under the most favourable circumstances lifted twenty-seven millions, which was the greatest duty ever performed till your petitioner's improvements were adopted, since which the greatest duty ever performed has been sixty-seven millions, being much more than double the former duty. That, prior to the invention of your petitioner's boiler, the most striking defect observable in every steam-engine was in the form of the boiler, which in shape resembled a tilted waggon, the fire being applied under it, and the whole being surrounded with mason work. That such shaped boilers were incapable of supporting steam of a high pressure or temperature,

and did not admit so much of the water to the action of the fire
as your petitioner's boiler does, and were also in other respects
attended with many disadvantages. That your petitioner's invention consists principally in introducing the fire into the midst of
the boiler, and in making the boiler of a cylindrical form, which
is the form best adapted for sustaining the pressure of high
steam.

" ' That the following very important advantages are derived
from this your petitioner's invention. This boiler does not
require half of the materials, nor does it occupy half the space
required for any other boiler. No mason work is necessary to
encircle the boiler.

" ' That, had it not been for this your petitioner's invention,
those late vast improvements which have been made in the use
of steam could not have taken place, inasmuch as none of the
old boilers could have withstood a pressure of above 6 lbs. to the
inch beyond the atmosphere, much less a pressure of 60 lbs. to
the inch, and is capable of standing a pressure of above 150 lbs.
to the inch.' "

Trevithick's retrospect views of 1828 are supported
by the letter of the late Michael Williams, M.P., the
most experienced of Cornish mine workers, but belonging to the eastern district that had been for many years
the users of the Watt engines in Cornwall.

" In reference to his great improvements in the steam-
engine, I have a more particular recollection, and can confidently affirm that he was the first to introduce the high-
pressure principle of working, thus establishing a way to the
present high state of efficiency of the steam-engine, and forming
a new era in the history of steam power. To the use of high-
pressure steam, in conjunction with the cylindrical boilers, also
invented by Mr. Trevithick, I have no hesitation in saying that
the greatly-increased duty of our Cornish pumping engines,
since the time of Watt, is mainly owing; and when it is recollected that the working power now attained amounts to double
or treble that of the old Boulton and Watt engine, it will be at
once seen that it is impossible to overestimate the benefit con-

ferred, either directly or indirectly, by the late Mr. Trevithick
on the mines of this county. I have often expressed my opinion
that he was at the same time the greatest and the worst-used
man in the county." [1]

The late Sir John Rennie and other scientific persons
were, about 1830, associated with Mr. Henwood [2] in
examining the work performed by Cornish pumping
engines: their reports are curtailed in the following
comments on Wheal Towan engine, similar to Trevithick's Dolcoath engine of 1816, except perhaps that
the last named was a little inferior in its detail movements, while much less care was taken
to avoid unnecessary loss of heat.

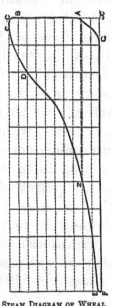

Mr. Henwood also gave indicator
diagrams of the expansion of the
steam, on one of which the writer
has marked ten horizontal lines, indicating the position of the piston at
each foot of its stroke, and ten longitudinal lines dividing the diameter of
the cylinder into tenths. The steam
pressure in the boiler was 46·8 lbs.
on the square inch above the atmosphere, or 4·68 lbs. for each of the
ten longitudinal line divisions. x to
c represents the top of the steam-
cylinder 80 inches diameter; x to
F the length of the cylinder for a
10-feet stroke of the piston. By the

STEAM DIAGRAM OF WHEAL
TOWAN PUMPING ENGINE,
ERECTED 1827.

time the piston had moved through one-twentieth of
its course, reaching c, the expansive working had com-

---

[1] See letter of Michael Williams, chap. xix.
[2] Henwood, 'Edinburgh Journal of Science,' 10.

menced; and when one-tenth of the stroke had been
run, half of a division was cut off, showing by the
curved indicator line the decrease in pressure of steam
to 44·46 lbs. The comparatively small passage through
the steam-valve not giving room for sufficient steam to
follow up the increasing speed of the piston, led to its
continued expansion in the cylinder, and by the time
the piston had moved 2 feet, reaching D, the steam
pressure was reduced by two divisions or 9·36 lbs.,
or a pressure of 37·44 lbs. on the piston; at this point
the steam-valve was closed, and the remaining four-
fifths of the stroke was performed by expansion; at
the fifth horizontal line, or middle of the stroke, only
three divisions of steam are left, giving a pressure of
14·04 lbs. to the inch; at the finish of the stroke
there is only half a division, from E to F, or 2·34 lbs.
of steam to the inch above the pressure of the atmosphere. On the return up-stroke of the piston, when it
had reached within a foot of the finish of its course at C,
the equilibrium valve closed, causing the enclosed steam
of 2·34 lbs. to the inch to be compressed at the finish
of the up-stroke shown by the curve G A to 9 36 lbs. on
the inch, equal to its pressure about the middle of the
down-stroke at N.

Trevithick's expansive engine therefore, commencing
its work with steam of 46·8 lbs. on the inch above the
atmosphere, only took a full supply from the boiler during
one-tenth of its stroke, and none after one-fifth had been
performed, while at the finish of the stroke it had about
the same pressure as Watt began with.

The *power* of the Watt low-pressure steam vacuum
pumping engine was increased by Trevithick from two
to three fold, and its economical duty in about the same
proportion; in other words, he increased the effective

power of the steam-engine two or three fold without additional consumption of coal.

In the Wheal Towan engine the steam-cylinder was 80 inches in diameter, with a 10-feet stroke. The shaft was 900 feet in depth ; the main pumps 16 inches in diameter ; the pump-rods were of wood, about 14 inches square, and weighed more than the column of water in the pipes. The boilers were Trevithick's cylindrical with internal tube, wholly of wrought iron. The cylinder and steam-pipes were surrounded with sawdust about 20 inches in thickness, as a non-conductor of heat. The upper surfaces of the boilers were covered with a layer of ashes for the same purpose. The duty performed was 86 58 millions of pounds of water, raised one foot high by the consumption of a bushel of coal weighing 84 lbs. The immense power and economy of this engine are best understood by its average labour costing only one farthing in coal for lifting 1000 tons one foot high.

At or about that time an old intimate of Trevithick's, Captain Nicholas Vivian, managed the mine, and Mr. Neville, a shareholder, also a user of steam-engines in Wales, observing the economical working of Wheal Towan high-pressure steam expansive engine, doing eighty-seven millions, requested its manager to examine colliery engines, all of which were of the low-pressure kind ; one of them was a Newcomen atmospheric, whose duty was six millions ; four or five others were Watt low-pressure steam vacuum engines, doing fourteen millions ; therefore the high-pressure steam-engine did six times as much work with a bucket of coal as the low-pressure steam vacuum, and fourteen times as much as the low-pressure steam atmospheric engine. Several competitive trials by the county engineers were pub-

lished about that time, in one of which, after a personal
examination of the engine, Mr. W. J. Henwood[1] and
others reported a duty of 92·6 millions with a 91-lb.
bushel of coal.[2]

Mr. Rennie had been a pupil, a fellow-worker with
low-pressure Watt, and while his son, Sir John Rennie,
was examining the high-pressure steam expansive
engine erected by Trevithick's pupil, Captain Samuel
Grose, under the management of Trevithick's friend,
Captain Nicholas Vivian, the latter was engaged in
reporting on certain low-pressure steam-engines in
Wales, one of which was a Newcomen's atmospheric,
probably the last of its race, whose principle of construction was a century old, working in company with
the Watt low-pressure steam vacuum engine, then half
a century old, the principles of both systems being on
their last legs, and under the care of Trevithick's
supporters.

During this jumble of engines, old and new, without
a clear comprehension of their differences in principle,
Trevithick, who had just returned from America, and
lived within a few miles of Wheal Towan, looked on
unconsulted and unconcerned on questions which in his
mind had been settled by him in Dolcoath fifteen or
twenty years before. The writer, during the Wheal
Towan controversy, was the daily companion of Trevithick, and made drawings of the engine at the works
of Harvey and Co., of Hayle, where it was constructed
about 1827.

Captain Samuel Grose's Wheal Towan engine was
in general character similar to his teacher's Dolcoath

[1] Address, Royal Institution of Cornwall, by W. J. Henwood, 1871.
[2] Trevithick calculated 84 lbs. to a bushel; Watt generally 112 lbs.; Lean
94 lbs., but latterly 112 lbs.

76-inch engine of 1816, working with about the same steam pressure and degree of expansion. The valves, gear, and nozzles were perhaps improved in detail; but the groundwork was unchanged. The first high-pressure steam Cornish pumping engine made in France was designed and superintended by the writer at the works of Messrs. Perrier, Edwards, and Chaper, at Pompe-à-feu, in Chaillot, a suburb of Paris. The principle was the same as the Dolcoath engine, and the detail differed but little from it or the Wheal Towan, except that its exterior was a little more artistic than its prototypes in Cornwall, in keeping with French requirements. It was built in 1836, within a few yards of the low-pressure steam pumping engine erected by Perrier and others in 1779, which still continued pumping water from the Seine for the supply of Paris. Stuart says, " An engine by Boulton and Watt was sent to France, and erected by M. Perrier at Chaillot, near Paris. The French engineer, Proney, with a detestable illiberality, attributes all the merit of the improvements in the Chaillot engine to his friend Perrier, the person who merely put together the pieces he had brought from Soho."[1]

The Perrier of 1779 was related to the Perrier of the Pompe-à-feu engine-building works of 1836, and his nephew took the Trevithick engine from Paris to a coal mine not far from Brussels, but not fully understanding the use of the balance-bob—the woodwork for which had not been completed in Paris, though all other parts had been fully erected—did not find it easy to manage the engine. The writer viewed Perrier's move as an infringement of the agreement between him and Edwards,

---

[1] Stuart ' On the Steam-Engine,' p. 141.

the partner of Perrier and Chaper, and therefore declined to take any further interest in the engine.

Mr. Edwards had before that been a partner with Woolf, in a small engineering works in Lambeth; and the writer had also before that been a pupil of Woolf's, in the works of Messrs. Harvey and Co., of Hayle.

The drawing of 'La Belle Machine' (Plate XIII.), of 1836, serves not only as a record of that time, but also in conjunction with the drawing of Dolcoath engine of 1816, enables an engineer to form a sufficiently correct idea of the Wheal Towan engine and boilers of 1827, which in effective duty is scarcely excelled by the best pumping engines of the present day.

The events connected with those Paris engines bring together the engineering works of Watt, Proney, Perrier, Trevithick, and Woolf, in the person of his once partner, Edwards. The writer, when constructing 'La Belle Machine,' had not the slightest knowledge of those links, and heard the name and repute of his engine by the following chance :—

In 1838 a passenger leaving the train of the Great Western Railway at Drayton Station, asked the writer's permission to walk on the line and examine its construction. During a short conversation he mentioned the having purchased at a sale in France the drawings of an engine known as 'La Belle Machine,' representing the Cornish high-pressure expansive steam pumping engine :—*a*, steam-cylinder, 48 inches in diameter, 8-feet stroke; *b*, steam-pipe from boiler; *c*, regulating steam-valve, double beat; *d*, regulating rod and handle for steam-valve; *e*, expansive steam-valve, double beat; *f*, balanced lever and rod for opening expansive valve; *g*, expansive clamp on plug-rod, with regulating rod and thumb-screws; *h*, cataract-rod for relieving expansive

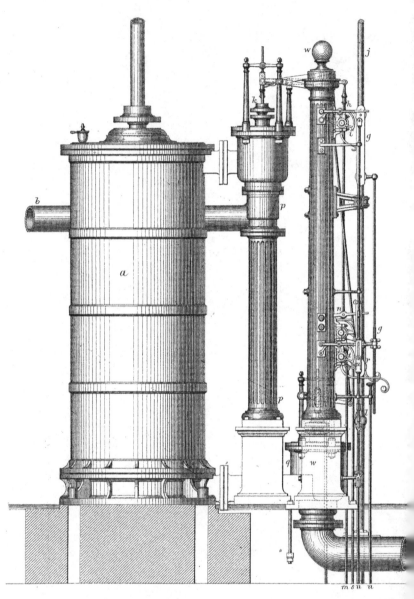

LA BELL

HIGH PRESSURE STEAM EXPANSI

London: E. & F. N. Sp

PLATE 13.

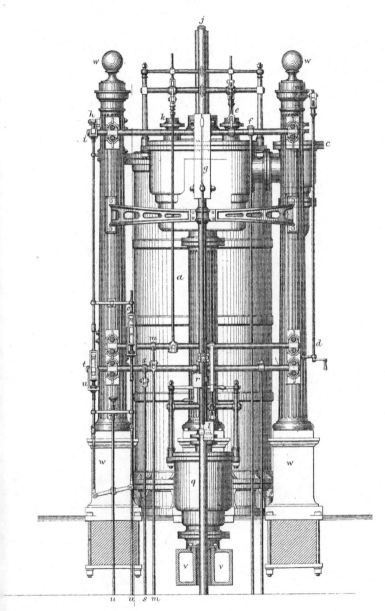

MACHINE.

PUMPING ENGINE. _ 1836.

Kell, Broˢ Lith. London.

valve-catch; *i*, quadrant relieving the catch; *j*, plug-rod; *k*, equilibrium valve, double beat; *l*, clamp in plug-rod to close equilibrium valve by its action on the handle; *m*, balanced lever and rod to open equilibrium valve; *n*, quadrant and catch relieving equilibrium valve by the action of cataract-rod; *o*, regulating slide on cataract-rod; *p*, equilibrium steam-pipe conveying steam from the top to the bottom of the piston; *q*, exhaust-valve, double beat; *r*, clamp on plug-rod, closing the exhaust-valve by its descent on the handle; *s*, balance lever and rod, opening exhaust-valve; *t*, quadrant and catch, relieving equilibrium valve by the action of cataract-rod; *u*, regulating slide on cataract-rod; *v*, exhaust-pipe to condenser; *w*, Y-posts for carrying the gear. The steam in the boiler was from 40 lbs. to 50 lbs. on the square inch above the atmosphere.

Lean states that had the pumping engines at work in Cornwall in 1835 remained unimproved since 1814, at which time they had benefited by three years of continuous improvement, a yearly additional expenditure of 80,000*l.* for coal would have been the consequence, and that the first step was Trevithick's expansive steam from the cylindrical tubular boiler, engines using such steam performing a duty three or four fold what Boulton and Watt had ever attained, or perhaps thought possible of attainment.[1] The birth of the idea of using expansive steam may in truth be traced back nearly one hundred years to the time of Newcomen's atmospheric engine, and the hope expressed in 1746 of a smaller boiler and more elastic steam[2] was partially realized in the engine and boiler of Trevithick, sen., in Bullan Garden in 1775, followed in 1780 by the competing

---

[1] See Lean's Historical Statement, p. 154; published 1839.
[2] See vol. i., p. 7.

engine erected by Watt in Dolcoath Mine, under Trevithick's management. Little further change was made until 1799, when the globular boiler and internal tube of Trevithick, jun., gave a second start to the use in large engines of more expansive steam; and even this partial move was the result of years of thought and practical experiment; for in 1792, when twenty-one years of age, he was the elected judge on a competitive trial between the Watt engine at Seal-hole, patented in 1782, and Hornblower's double-cylinder engine at Tin Croft. Each engine performed a duty of ten millions, both of them were called expansive, while in fact neither of them were so, for the pressure of the steam in the boiler did not admit of it. As Lean says, "As the steam used was raised but little above the pressure of the atmosphere, it was found that the power gained did not compensate for the inconvenience of a more complicated and more expensive machine." Or, as Watt said to Robert Hart, "We resolved to give up the expansion of the steam until we could get men that could work it," as he found it more costly than profitable. Again in 1798, Trevithick's own writing records his experiment in Dolcoath between the Bullan Garden 45-inch atmospheric engine and the Watt 63-inch great double-acting engine, when the latter did sixteen millions to ten millions by the atmospheric. At that very time he was constructing his high-pressure steam portable engines, and in the following year, after seven years of most active experience, prompted by the Watt lawsuit against Cornish engineers, he in 1799 gave the beaten 45-inch engine steam of a higher pressure from the stronger globular boiler. People, following the ideas of Watt, were still afraid of Trevithick's plans, distinctly laid down in his letters of

1806, recommending a cylindrical boiler for the Dolcoath pumping engine, because similar boilers giving
steam to his whim-engines have enabled them to beat
the Watt whims. This continued until 1810, when the
greatly - increased power and economy of the high-
pressure expansive steam pumping engine at Wheal
Prosper caused the neighbouring Dolcoath in 1811 to
give Trevithick's plans free scope. The long smoulder-
ing rivalry between low and high pressure, on the eve
of the final discomfiture of the former, burst forth in
loud words and evil prognostications, causing the
mining interest of Cornwall to appoint an examiner
who should publish monthly the duty performed by the
various pumping engines, the first of which appeared in
the autumn of 1811, when Trevithick was building his
boilers in Dolcoath, and preparing the engines, as far as
was possible, to submit to strong steam.   By expansive
valves and suitable gear, balance of power between the
engine and the pump-work necessitating balance-bobs,
strengthening the pit-work to bear the more powerful
and sudden movement, and fifty other things, which
we know must have presented themselves in such
work, occupied the greater part of Trevithick's time
from 1811 to 1814.   That first report enumerates
twelve pumping engines, probably all of them Watt
engines, averaging a duty of seventeen millions.

We have before traced the rapid and immense
increase in the power and in the duty of Cornish
pumping engines from 1811, and it may be taken as
comparatively true in the larger sense applying to the
improvement of the steam-engine everywhere.

Dolcoath Mine, one hundred years ago, under the
management of Trevithick, sen., followed by his son as
the strong-steam engineer, and by his grandson as one

of the committee of management in these modern times, has served during that long period to illustrate the progress of the steam-engine, and still in active operation, was thus spoken of in 'The Times' of Dec. 18th, 1871 :—" This old and extraordinary mine is now raising about 100 tons of tin every month, worth from 8000*l.* to 9000*l.*"

CHAPTER XXI.

ENGINES FOR SOUTH AMERICA.

[Rough draft.]

" SIR,   "CAMBORNE, *May 20th*, 1813.

"Yours of the 7th inst. I should have answered by
return, as requested; but an unexpected circumstance prevented my being at Swansea as early as proposed, which, as it
happens, best suits your purpose as well as my own. I shall
not be able to be there within twenty days from this time, of
which I will give you timely notice. I hope before that time
Mrs. Rastrick will be safe out of the straw. I have been detained
in consequence of a strange gentleman calling on me, who
arrived at Falmouth about ten days since, from Lima, in South
America, for the sole purpose of taking out steam-engines,
pumps, and sundry other mining materials to the gold and
silver mines of Mexico and Peru. He was recommended to me
to furnish him with mining utensils and mining information.
He was six months on his passage, which did not agree with
his health, and has kept his bed ever since he came on shore;
but is now much recovered, and hopes to be able to go down in
the Cornish mines with me in a few days. I have already an
order from him for six engines, which is but a very small part
of what he wants. I am making drawings for you, and intend
to be with you as soon as they are finished. Money is very
plentiful with him, and if you will engage to finish a certain
quantity of work by a given time, you may have the money
before you begin the job. The West India engine will suit his
purpose. I shall have a great deal of business to do with you
when we meet. In the meantime please to forward the thrashing engines to Cornwall as quickly as possible. The engine for
Plymouth will be put to break the ground as soon as I can find

time to go up there.   Please to say when and by what ship I
shall have the small engines.

" I remain, Sir,

" Your very obedient servant,

" R. T.

" To Mr. John U. Rastrick,
  " Bridgenorth, Shropshire.

"The copper mine mentioned in my last is improving very
fast."

The strange gentleman reterred to was Don Francisco
Uville, a person of great influence in Lima, who a year
or two before had travelled from Peru to England and
back, in search of steam-engines to pump water from
the ancient gold and silver mines then flooded and idle.
Boulton and Watt, at Soho, on being consulted, dis-
couraged the attempt, because of the difficulty of conveying heavy machinery over mountain pathways, and
also because their low-pressure vacuum engine, using
steam but slightly above atmospheric pressure, would
be much less effective in the comparatively light atmosphere on the high summits of the Cordillera Mountains
than in England.   Uville, who had heard of the
wonderful ability of English engineers to construct
steam pumping engines, was utterly downhearted at
this decision of the great Soho engineers, and while
dejectedly wandering through the streets of London,
unconsciously gazed into the shop window of Mr. Roland
in Fitzroy Square, near the spot on which Trevithick
had run his railway locomotive three years before.[1]
Rumour of passed events may have led him to visit the
ground on which had worked a new kind of steam-
engine.   His searching glance discovered among numerous articles for sale, an unknown form that might be

---

[1] See vol. i., p. 194.

the talisman he had travelled thousands of miles in search of. The shopkeeper informed him that it was a model of Richard Trevithick's high-pressure steam-engine, which worked without condensing water, or vacuum. If what he heard was true, why should it not work equally well in the light atmosphere of the mines? The great engineer at Soho might be in error or ignorance. The experiment, as a last resource, was worth making. He would pay the 20l. for the model, carry it to the mines of Cerro de Pasco, in the high mountains above Lima, where, if it worked as well as it did in London, the rich mines of Peru would again reveal their long-hidden treasure. The model was conveyed by ship to Lima, and then on a mule up the narrow precipitous ascents to Cerro de Pasco, over mountains more than 20,000 feet high. Fire was placed in the small boiler as he had seen it done in London, and with the same result, to the great joy of Uville, who determined to revisit England in search of the inventor of this new and wonderful power. On his return voyage, when rounding Cape Horn, bets were made on the chances of his finding the man who had invented the high-pressure steam-puffer engine,[1] and of his being able to persuade such a person to make the required engines and accompany them to Peru. Such gloomy forebodings ended in an attack of brain fever. The vessel touched at Jamaica, where Uville was landed. On recovering health and strength he embarked for England in one of the packet-ships, and during the voyage still spoke of the object of his search. A fellow-traveller, called Captain Teague, rejoiced him by saying, " I know all about it; it is the easiest thing in the world. The inventor of your high-

---

[1] See London locomotive, vol. i., p. 198.

pressure steam-engine is a cousin of mine, living within a few miles of Falmouth, the port we are bound for." On landing, Uville, still weak and obliged to keep his bed, was told that Trevithick, the engineer, lived in London, and was constructing the Thames Tunnel; but further inquiry showed that he also had suffered from brain fever, and had just returned to Penponds, only a few miles from Falmouth. On the 10th of May, 1813, a letter reached Trevithick, requesting him to visit the sick Uville, and in a fortnight from that time the engineer had mastered the requirements of the Peruvian mines, and had designed and made arrangements for the supply of six pumping engines, together with the pumps and all things necessary for the underground workings; the whole to be delivered in four months.

[Rough draft.]

" SIR,　　　　　　　　　　　　　" CAMBORNE, *May 22nd*, 1813.

"I have engaged to get six engines, with pit-work, &c., to send abroad. A great part of the wrought-iron work and the boilers I have arranged for in Cornwall. These engines will be high-pressure engines, because the place they are for has a very deep adit driven into the mountain; and lifting condensing water to the surface would be a greater load than the whole of the work under the adit level.

"I call a set of work, a 24-inch cylinder single engine, 6-feet stroke, piston, cylinder bottom, single nozzle, with two 5-inch valves and perpendicular pipe; no cylinder top; the piston-rod not to be turned; 3-inch safety-valve, fire-door, two small Y shafts and gear-handles, &c.; a good strong winch set in a broadish frame, such as is often used on quays or in quarries, 25 fathoms of 12-inch pumps, a 12-inch plunger, an 11-inch working barrel, clack-seat and wind-bore, with brass boshes and clacks, a force-pump for the boiler, and 10 fathoms of 3-inch pipes to carry the water to and from the engines. I have engaged to supply six full sets of the above-mentioned materials.

" All these castings must be delivered in Cornwall in four months from the time the orders are given; therefore, if you take the job, or any part of it, you must enter into an engage ment to fulfil it in the time. As there ought not to be a moment lost, I wish you to answer me immediately in what time you will deliver those materials in Cornwall; or otherways, what part of them you can execute in the time.

" I am making the drawing, which will be ready before I can receive your answer. For whatever part of the job you may engage I will lodge the money to pay for the whole in Mr. Fox's hands, which will then be paid for before you begin the work, as soon as you execute the agreement.

" R. T.

" Mr. PENGILLY, *Neath Abbey, South Wales*."

It is an odd coincidence that while writing of the events of fifty-eight years ago, pumping engines are being sent to those same mines with the steam-cylinder in twenty-two pieces, no piece to weigh more than 300 lbs.—a facility in mechanical arrangements not enjoyed by Trevithick—having Trevithick's high-pressure boilers, giving steam of 50 lbs. on the inch.[1]

[Rough draft.]

" SIR,   " CAMBORNE, *June 2nd*, 1813.

" I drop you this note just to inform you that I have begun your job. Yesterday I engaged a great many smiths and boiler-builders, who set to work this morning. I have also engaged all the boiler-plates in the county, which will be sent to-day to the different workmen. The master-smiths that I have engaged are the best in the kingdom. I have obligated them to put the best quality of iron, and to be delivered at Falmouth within four months. I have been obliged to give them a greater price than I expected, otherwise they would not turn aside their usual business employment for a short job of four months.

" Mr. Teague is with me, and one other, assisting about the

---

[1] Made by Harvey and Co., Hayle, 1870.

drawings. If you call at Camborne about Friday, shall be able
to show you the designs. The drawings for the castings will be
sent to the iron-founders by the end of this week; and by the
end of next week shall have the whole of the different tradesmen in full employ. If you wish to have a greater quantity of
machinery ready by the end of September, there ought to be as
little time as possible lost in giving your orders. I can get you
double the quantity, provided you give the orders in time.

"As soon as it is convenient to you to arrange the payments
I would thank you to inform me, because we find in practice
that the best way to make a labouring machine turn quickly on
its centres, is to keep them well oiled.

" R. T.

"F. Uville, Esq., Mr. Hooper's, *Falmouth.*

" N.B.—If you intend to be at Camborne, please to drop me
a note by post, and I will be at home."

In all Trevithick's moves there was a scramble for
money, in which he invariably came worst off. He
could give a good hint that working centres would not
turn well without the essential oil; but he failed to
apply the principle to himself. Liberal words and
golden prospects carried him off at once; and before
Uville was strong enough to visit the Cornish mines
and to fully explain what he wanted, the machinery
was being made, though at that same time the thrashing and ploughing engines, and the locomotive and
rock-boring engine, and the great fight with Watt at
Dolcoath, were in progress.

[Rough draft.]

"Mr. Rastrick,   "Camborne, *June 8th*, 1813.

"Sir,—Enclosed I send to you a drawing for a set of
pumps for one of the engines for South America, with a
drawing for a part of the castings for one of the boilers, for you
to make a beginning. The drawings for the engines I will send
in a few days. The Spanish gentleman who is now gone to

London to arrange his money concerns, will be down again in about ten or twelve days, and then we shall both call at Bridgenorth, and bring with us the engagement for you to sign, for the performance of such quantities of work as you can execute in four months.

"I have made arrangements with the smiths and boiler-builders here, to weigh and pay at the end of every week. The regulation of your payment is left to you to point out in any way you please. As time is of the greatest consequence, I hope you will set to work immediately.

"The reason for making the pumps so short, is on account of the extreme badness of the roads over the mountains, where these engines are to be conveyed, it being almost impossible to carry above five hundredweight in one piece. The West India engine is sold to send to Lima, but not to be conveyed over the mountains. I shall also bring drawings with me for one or two winding engines for the same place. Please write to me by return of post.

"R. T."

[Rough draft.]

"CAMBORNE, near TRURO, *June 11th*, 1813.

"MR. FRANCIS UVILLE,

"at MESSRS. CAMPBELL AND Co.'s, London.

"Sir,—I have your favour of the 9th instant, respecting the weight of the largest parts of the engines. I will take care to reduce the weight if possible, so as to be carried on the backs of mules.

"By the time I receive your letter I shall have arranged the whole of the engine business, and intend to go immediately to Wales and Shropshire, to get the engagements executed for the performance of the work by the time proposed. I shall write to you again before I leave home, and as soon as I arrive in Wales will also write to you. I shall not stay in Wales above two days, but go to Bridgenorth in Shropshire, where I hope to have the pleasure of meeting you, as it will only be about twelve hours' ride out of your road to Cornwall.

"In the North I shall introduce you to the sight of a great deal of mining and machinery, and in about ten days from the time you arrive at Bridgenorth, shall be able to accomplish the business so as to return again to Cornwall.

"I would thank you to inform me as early as you can, of the number of engines you intend to get executed by the proposed time, because when I am in the North I shall be able to arrange with the founders accordingly. The smiths are all at work for you.

"R. T."

[Rough draft.]

"Mr. Uville,   "Cornwall, Camborne, *June 19th*, 1813.

"Sir,—Your favour of the 9th instant, dated from Falmouth, I received, and in return wrote to you immediately—directed for you at Messrs. Campbell and Co.'s, London. As you said in your last letter, that immediately on your arrival in town you would write to me, I have expected every post since last Tuesday would have brought me a letter; but as I have not received it according to your promise, I am fearful that your letter may be unexpectedly detained, especially as you told me the last time I saw you at Falmouth, that you would enclose me a bank post bill. All the founders and other tradesmen are in full employ on your engines.

"I intended to have left Cornwall for Wales and Shropshire by this time, with the founders' articles for execution; but being disappointed in not hearing from you, agreeable to our appointment, I shall delay it until I hear from you, which I must request you to have the goodness to do by return of post, because those delays make very much against the execution of your work; and as time is of so great a consequence to you, I hope you will not lose a moment in writing and giving me the necessary instructions, with a few drops of that essential oil that you proposed sending me on your arrival in town.

"R. T."

The sugar rolling-mill engine that had been made for the West Indies so pleased Uville that he purchased it at once, intending it for the Mint at Lima. He also ordered one or two winding engines, in addition to the pumping engines. Trevithick had arranged that no piece should exceed 560 lbs. in weight. Then came Uville's order, "if possible to be reduced so as to be carried on the backs of mules." Since that time the path on the mountains has been improved, yet the present limit of weight is 300 lbs. The absence of the promised bank post bill was another difficulty.

[Rough draft.]

"CAMBORNE, *June 23rd*, 1813.

"MR. FRANCIS UVILLE,

"at MESSRS. CAMPBELL AND Co.'s, Park Buildings, London.

"Sir,—Your favour of the 19th instant came safe to hand.

"I was in hopes that I should have found a remittance enclosed. All the tradesmen that I have employed on your work were to have been paid every Saturday, and I made my arrangement with you accordingly. Unless this mode of proceeding is followed up, you cannot get your work done in any reasonable time, especially as you are an entire stranger. For my own part I have placed the greatest confidence in your honour, with which I am fully satisfied.

"But I have to get this work from a great number of different tradesmen, and must make regular payments agreeable with my engagements with them. As the articles are about to be executed by different tradesmen, regular weekly payments ought to be established, of which I informed you before the work began.

"I am ready for my journey to Wales and Shropshire, but cannot proceed with further engagements until I hear again from you. I have placed the fullest confidence in your word, a proof of which you have in the great exertion I have made to get the work done; but unless you in return place some confidence in me, or any other engineer that you may employ, a work of this magnitude cannot be carried on with promptitude.

"As the whole of the work in my part has been put into immediate operation, it would be a very serious loss both of money and time to discharge the hands. I hope you will fully consider this business, and must beg you will have the goodness to write to me by return of post. On receiving the needful from you I shall leave Cornwall for Wales and Shropshire.

"R. T."

Trevithick for once in his life was wise, and would not start on his journey to Bridgenorth until the money had reached him. This prudent resolve was soon forgotten in the love of making the steam-engine useful; and as such creations in his hands grew into shape and size before other men would have got through preliminary discussions, pecuniary difficulties sprang up, as mushrooms do in a night.

[Rough draft.]

" CAMBORNE, September 4th, 1813.

" MESSRS. HAZELDINE, RASTRICK, AND CO.,

"Gentlemen,—Enclosed you have three of Mr. Uville's drafts, value one hundred and fifty pounds.

"I should have sent it in one draft, but had not a suitable stamp. The castings, pipes, ale, &c., arrived safely. I hope that all the boilers and wrought-iron work will be finished by the end of this month, and shipped off for London. Immediately after Mr. Uville and I shall leave Cornwall for Bridgenorth on our journey to town. We are both very anxious to see the 'Sanspareil' engine at work, and hope you will have it ready by that time. I have received orders from different persons since I have been here, for steam-engines for the West Indies, and must, if possible, have three ready early in November, as the ships sail then that will take them.

"I wish you would say in your next if this can be done in time, because these persons are very extensive agents for the planters, and are extremely anxious to generally adopt them in the West Indies.

"We find from your letter that you are getting on pretty fairly with Uville's work.

"I remain,

"Your very humble servant,

"RICHARD TREVITHICK."

[Rough draft.]

" GENTLEMEN,   " CAMBORNE, *September 7th*, 1813.

" After writing to you on Sunday last, Mr. Uville received letters from Cadiz, from the Spanish Government, informing him that there was a line-of-battle ship there that should take the engines to Lima. Now as this ship is detained for this purpose, all possible dispatch must be made to get the whole of the materials shipped as early as possible for Cadiz. I am pushing the smiths as hard as possible, and you must do the same at your works, that the greatest dispatch may be made. I am ordered by Mr. Uville to request you to get one water-engine, pumps, &c., complete, one winding engine, winding apparatus, &c., complete, and one crushing apparatus, complete, in addition to the former order. I wish you would also get on as fast as possible with the new engine, but do not let this engine prevent the getting forward the work for Lima.

" I wish to have made apparatus to work expansively, and also a temporary water-pump, to load the engine, so as to prove its duty by the consumption of coal.

" If the jobs are not completed by our arrival, you need not expect any rest until its completion. Your answer will oblige,

" R. T."

" MESSRS. HAZELDINE, RASTRICK, AND Co."

The money difficulty was for a time surmounted, with a prospect of the completion and shipment of the work for London within four months of the giving of the order; and the Spanish Government proposed that a line-of-battle ship should take the engines to Lima from Cadiz. An order was given for another pumping engine and another winding engine, to be provided with gear for working expansively, and a temporary water-pump, that in case of need the amount of work the engines could do with a given amount of coal might be tested. A crushing machine, now called " quartz-crusher," also formed part of this additional order.

The new engine, which he hoped they would get on with, was probably the steam locomotive plough then being constructed at Bridgenorth.

[Rough draft.]

" GENTLEMEN,   "CAMBORNE, *September 22nd*, 1813.

"I have your favour of the 14th instant, and hope to find you as forward on your job on our arrival at Bridgenorth as you state.   I expect all the boiler and smith work will be shipped for London early in October; we shall then leave Cornwall for your works, at which time you will be very much annoyed with our company, unless we find your assertions grounded on facts.   Enclosed I send you Mr. Uville's draft for 150*l.*   Your receipt for the draft enclosed in my letter of the 16th instant has not yet arrived.

"I hope you will also have all the apparatus ready to try the new engine; Mr. Uville is very anxious to take the first of these new engines with him.   When you send a receipt for the enclosed, please to say what state of forwardness the whole of our work is in, and do not neglect a moment to get the whole executed with all possible dispatch.

"Nothing short of a want of cast iron will confine our friend in England one day after the end of this month.

                    "I am, Gentlemen,
                         "Your very humble servant,
                              "RICHARD TREVITHICK.
"MESSRS. HAZELDINE, RASTRICK, AND CO."

It seems probable that in 1813 a railway locomotive, with apparatus for rock boring, and steam-crane, was made for South America as the forerunner of the 'Sanspareil' of 1829.

[Rough draft.]

" GENTLEMEN,   "CAMBORNE, *October 1st*, 1813.

"I received your favour of the 27th last evening, and now enclose you another draft of Mr. Uville's for 150*l.*   We shall wait impatiently for your next letter to know when you will finish. Mind, this is the 1st of October, and agreeable to promise the time is up. Mr. Uville wishes you to cast sixty carriage-wheels for him, 11 inches in diameter from out to out, and to weigh about 20 lbs.; cast them of strong iron, and of a strong pattern, to take a 1½-inch axle by 2½ inches deep in the hole; also cast four plunger-pistons 11 inches diameter to suit the 11-inch working barrels, provided it should be used for the purpose of a plunger. They must be in every respect the same as the 14-inch plunger-pistons, only 3 inches less in diameter.

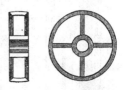

"Soon after the receipt of your next letter you may expect to see us, as a vessel has been engaged to take all the boilers and smith work on board to-morrow week for London.

"I remain, Sir,

"Your humble servant,

"RICHARD TREVITHICK.

"MESSRS. HAZELDINE, RASTRICK, AND Co."

Probably those cast-iron wheels were ordered with a view to steam locomotion in the Cordilleras. An engine is described in the invoice as having chimney, axles, carriage-wheels, &c.

[Rough draft].

"GENTLEMEN,   "CAMBORNE, *October 11th*, 1813.

"On making the drawings of the engine with the winding and crushing apparatus, when at work I find that if there is no crank, but the sweep rod is connected to a pin in the arm of the fly-wheel; in that case the fly-wheel will cut off the engineer from getting at the cock; but if the sweep is connected to a crank, then there will be sufficient room. The copy of materials taken from your books and given to Mr. Uville does not say in which way it was intended. I send you a sketch how it will stand worked by a pin in the fly-wheel, and also if worked by a crank over the cylinder, with the fly-wheel outside the wood partition of the house. If you have cast all the parts for the winding engine, you should try to alter it, having the fly-wheel outside the wall of the house, and a crank for the

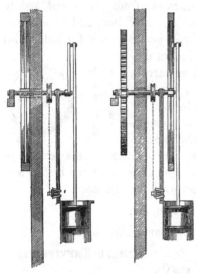

inside end of the shaft. The fly-wheel shaft will be nearly the same length both ways, only it must be long enough for the fly-wheel to pass between the wood partition and the 4-feet cog-wheel. The centre of the winding cylinder will be 17 inches from the outside of the wood end of the house, against which the fly-wheel ought to run. I have received your favour of the 5th instant, and have enclosed, agreeably to your request, a draft of Mr. Uville's for 800*l.*, which will be the

WINDING ENGINE FOR SOUTH AMERICA.

last from Cornwall. All I have to say is, you have taken longer time for the completion of your work than you first proposed, which has made Mr. Uville apprehensive that it will be the means of his losing the Spanish ship promised him to take the engines. He desires me to inform you that he has complied with this advance on purpose to enable you to push your work with the utmost exertion.

"Please to inform us the precise time we must quit Cornwall for Bridgenorth; we now wait entirely on you without any other thing to engage us. I fear Mr. Rastrick being so much from home will impede our job. If we miss this ship it will certainly make much against us all, losing three or four months in getting a South Sea whaler, and having the engine in a vessel not able to defend herself against an enemy, and having to pay 15 or 20 per cent. insurance, and prevent our getting other orders for another set of engines, and if taken by the enemy perhaps altogether damn the undertaking. Therefore I would have you to well consider the great inconveniences attending delay.

" I think I need not say much more to you on this head, as you ought to feel more for your own interest than I can scribble to you on paper.

"Yours, &c.,

"R. TREVITHICK.

"MESSRS. HAZELDINE, RASTRICK, AND CO."

This rough hand-sketch and letter fully describing his requirements, is an illustration of the facility with which Trevithick designed his engines and made known his wishes to others.

[Rough draft.]

"GENTLEMEN, "CAMBORNE, *October* 23rd, 1813.

"Mr. Uville wishes everything to be sent off as soon as finished, except the rolling engine, which is to remain until he arrives. We intend to leave Cornwall for Bridgenorth on Monday, November 1st. You may expect to see us three days after that date. The wheels ordered for the carriages are to run on the ground and not on railroads. Mr. Uville now wishes to have seventy-two instead of sixty as ordered before.

"I remain, Gentlemen,

"Yours, &c.,

"R. TREVITHICK.

"MESSRS. HAZELDINE AND CO."

The last-named engine was intended for the coinage operations in the Mint at Lima. The use of railway locomotion had been under discussion with the engine builders, and probably those particular carriage-wheels were ordered in the hope that the portable engine built for conveying itself from place to place in the sugar plantations of Jamaica, would in the Cordilleras be made to draw waggons on common roads.

The hand sketch of the winding engine in the letter

of the 11th October, was to correct an error in an order hastily given a month before; when, to save time, outline instructions for this complicated work were hurriedly sent to the manufacturer, that a commencement might be made while the more perfect detail drawings were being completed; the first-proposed position of the fly-wheel would prevent the engineman from conveniently reaching the four-way cock; Trevithick therefore suggested that the fly-wheel should be moved to the outside of the house, and a crank placed on the end of the driving shaft in lieu of the crank-pin in an arm of the fly-wheel. The sketch illustrating this change makes us fully acquainted with the kind of winding high-pressure steam-puffer engines of 8-horse power, with open-top cylinders of 12 inches in diameter and about 3 feet 6 inch stroke, sent to Peru in 1814. Steam, of 30 lbs. to the inch above the pressure of the atmosphere, was admitted under the bottom of the piston by a cock moved by an eccentric on the fly-wheel shaft; the gradual closing of the cock reduced the supply of steam when about one-third of the stroke had been made, wholly cutting it off some time before its completion, making it a high-pressure steam expansive engine. The movement of the cock then turned the steam from under the piston into the chimney blast-pipe, and the down-stroke was performed by the weight of the descending piston, made more than usually deep and heavy to prevent the tendency to twist in the cylinder from the angle of the jointed connecting rod, and also by the momentum of the fly-wheel and its balance-weight, moving at a speed of thirty strokes a minute. Its boiler was the Trevithick wrought-iron cylindrical, with internal tube and fire-place, but so arranged that if necessary the fire could be placed in brick flues under the boiler, returning through the tube.

The cylinder for the winding engine was probably fixed in the boiler, costing, with whim-barrel and winding apparatus complete and ready for work, 210*l*. Does the reader ask, Did so cheap an engine ever work? Or perhaps his knowledge of engineering gives rise to the question, How did it work? for it looks like a Newcomen of just exactly a hundred years before, only it needs no injection water or great main beam; and certainly it is not a Watt, for it has neither air-pump nor condenser, nor vacuum, nor cylinder-cover, nor parallel motion, nor any other thing like Watt invented; but it has high-pressure steam, which he disapproved of, and it really worked thousands of miles away, where there were no mechanics to keep it in order, and on mountains so difficult of access, and in so light an atmosphere, that Watt, who had the first chance of supplying steam-engines to the New World, declared it to be impossible. The pumping engines are described in Trevithick's note of 22nd May. They also were high-pressure puffer-engines with open-top cylinder, 24 inches in diameter, 6-feet stroke, with a cross-head working in guides, and side rods connecting to the pump-rods. Two valves turned the steam on and off from under the piston, with the ordinary gear and handles. The boiler was similar to that for the winding engine, but larger, and had not the cylinder fixed in it; a balance-beam regulated the movements, as it had no great main beam, and differed from ordinary engines just as the winding engine did. The power was 33 horses, and with an 11-inch pump barrel, 150 feet of 11-inch pumps, a winch, and all apparatus necessary for draining the mine, the cost was but 1400*l*.

[Rough draft.]

"PLOUGH INN, BLACKWALL, *December 28th*, 1813.
"MR. RASTRICK,

"Sir,—I am requested by Mr. Uville to write to you, to push the boilers as fast as possible. A ship will sail for the South Sea fishery in about five weeks, and will engage to take the whole of the engines. We have not finally closed with her, because we cannot state the exact time until we hear from you. You must not lose a moment in sending the boiler to town. I should have gone to Cornwall before this, but have been detained, getting a ship; and I do not like leaving until my agreements are executed, which cannot be done until the beginning of next week.

"I have been obliged to have all the transactions between the mines, and the Spanish Government, and Mr. Uville, translated into English, before the outlines of an agreement could be drawn up, which has been a most tedious job.

"Most of the people have been out of town, and those that were not would do no business in the Christmas, which has occasioned a loss of near ten days.

"As soon as the agreements are executed, I will immediately send to you money from this place. I have been kept so long here, that it will not be worth returning to Cornwall until after Mr. Uville sails. I shall be at Bridgenorth in about ten days, and will remain until the work is finished. Write how the work is getting on, and what state the winding engine is in.

"Yours, &c.,

"RD. TREVITHICK."

[Rough draft.]

"DEAR SIR,   "CAMBORNE, *March 4th*, 1814.

"Your favour of the 23rd February was sent to me from Bridgenorth. I have also received your favour of the 1st instant, and will attend to the drawings you mention, and be prepared to meet you as early as you please, only give me as much notice as you can.

"I hope by this time that Mr. Page has done something

toward the needful, to be at your service. I have, agreeably with your letter this day, desired Capt. Thomas Trevarthen to hold himself in readiness for London about the end of this month. I have not yet seen Bull. I wish you to write me if I am to give him notice also to hold himself in readiness for town. I fear that those two persons will not be sufficient to conduct the work with speed, especially if Capt. Trevarthen should be unwell; he is a good miner and pitman, and could assist in fixing the engines. Bull can only act as an assistant to an engineer, therefore neither of them can take the sole direction of the work.

" There will be those four large boilers to be put together on the spot, which neither of those persons know but little about. I think it would take a great charge and care from your mind to have a third person with you that could go through the whole of the undertaking, especially as the distance from England is so great. This undertaking of such immense magnitude and value ought not to depend solely on your own health, as neither of the other two could get on without your assistance in laying down and planning the outline of the whole of the work belonging to the machinery. If any one of the parts should be lost or broken, it would require some ability in that country to contrive a substitute. The expense of a third able man might prevent much loss of time and difficulty, and would not be an object in a business of such a scale as you have commenced with.

" I recommend a third person, that you might count on a speedy and effectual start. Even in this kingdom, where machinery is so well understood, I have known several good undertakings fail, from not employing at first an experienced engineer to conduct the work; which I am doubtful would be the case at Pascoe, if you were not able to attend yourself to the erection, and do not take a person with you for that purpose. I beg your pardon for thus attempting to recommend to you a third person to go out; but I think a work of this magnitude, where expedition is important, ought not to rest on the health of one man, especially under a changeable climate. Please to consult your friends, and give me your opinion on it in your next.

"My health is much improved; my wife desires her best respects, and thanks for your present. Please to write soon.

"Yours, &c.,

"RD. TREVITHICK.

"MR. UVILLE, 12, *East Stien, London.*"

[Rough draft.]

"MR. PAGE,   "CAMBORNE, *March 8th,* 1814.

"Sir,—Yesterday Mr. Joseph Edwards, of Truro, informed me that Teague had given notice of trial, and that the case would come on at the Assize on the 26th, and requested me to desire you to write to him immediately, and give him the whole of the transaction relative to Mr. Uville's arrest in London.

"He also wishes that some attention had been paid to the threat that Mr. Uville received from Teague's so-called friend, so as to ascertain whether it came direct from him, which he thought would have some weight in court. I shall attend to give evidence at the Assize with Mr. Edwards. I shall anxiously await a reply to my last. How does Harvey's business get on?

"My respects to Mr. Day, and shall be very glad to find him recovering his health as fast as I am. A crust of bread and clear air are far preferable to luxuries enveloped in clouds of smoke and heaps of filth.

"Your obedient servant,

"RD. TREVITHICK.

"P.S.—I hear that Teague is still in London, and that his furniture is removed to his friend's house, to save it from the hands of surrounding evil spirits."

Trevithick showed no undue amount of discontent on discovering that Uville had led him into pecuniary difficulties, and even his tendency to interfere in engineering matters was not hastily resented.

In December, 1813, while in London, arranging for a vessel to convey the engines to Lima, and also to

secure written agreements with Uville, who expected
to leave England in a week or two, the going into the
documents made known many weak points, one of them
being shortness of money. The expected week or two
had lengthened out to three months, and Uville was
still in London, and Capt. Thomas Trevarthen and Bull
were to be there, ready to start, about the middle of
March, 1814. Four large boilers, in pieces, were to
go for the pumping engines, to be put together in
the mines; and Trevithick strongly recommended the
sending a third man, to take general charge of the
practical work, which Mr. Uville thought he himself
could manage.

Page and Day were lawyers, who drew up very long
documents. Money to pay expenses was raised by the
sale of shares in a company formed by Uville without
sufficient authority, and Page was to go to the mines to
look after his own and the English shareholders' interests; between them Uville was arrested, apparently
for some trifle.

[Rough draft.]

" MR. UVILLE,   "CAMBORNE, *March* 15*th*, 1814.

" Sir,—I shall write to him again by this post, and push
him to send down the transfer of my shares, already agreed on,
for my execution, and hope I shall be able to meet Messrs.
Hazeldine and Co.'s demand before it will be due. The young
man Bull has been with me. I told him I expected that
you intended to take him with you, and Capt. Trevarthen is
making preparation for going. I am glad you intend to take a
third person with you. I have not thought or said anything to
anyone about this business. Mr. Vivian informed me that, from
the conversation he had with you on the subject, he had
expected to hear from you. I can answer for Mr. Vivian's
honesty, ability, and pleasant behaviour, and he is a person
very suitable for the engagement, only that one failing of

making too free with an evening glass, which you were not unacquainted with while in Cornwall at Dolcoath Mine. I do not like to take an active part in this business, because if any accident should happen to him, my sister or his family might charge me with being accessory to his going; therefore I must beg to be exempt from taking any part in this engagement.

" I remain, Sir, yours,

" Rd. Trevithick."

[Rough draft.]

" Mr. Page,   "Camborne, *April 9th*, 1814.

" Sir,—I have your favour of the 5th instant. I intend to be in town on Sunday week, but this need not prevent their writing to me here; and both you and they may still be doing your best towards disposing of shares.

" Your obedient servant,

" Rd. Trevithick."

[Rough draft.]

" Mr. Uville,   "Camborne, *April 9th*, 1814.

" Sir,—I intend to be in London on Sunday, the 17th, and shall call immediately on this person for money, which shall be at your service. Wheal Alfred and Wheal Prosper agents wish you a prosperous voyage, and success in your mines.

" I remain, Sir,

" Your obedient servant,

" Rd. Trevithick."

Trevithick was now embarked with a crew of speculators, and in payment for his services was made a partner, and sold a portion of his shares to pay for the engines which Uville had ordered.

Henry Vivian, his brother-in-law, and the brother of his late partner Andrew Vivian, wished to be the third person engaged to go with the machinery to America.

Trevithick spoke of his honesty and ability, but declined, on account of the family relationship, to take any part in the appointment.

The two notes on the 9th April, 1814, close the correspondence. Page was busy selling shares to raise money, and Trevithick was to get some money, which was to be at the service of Uville.

The delay between this period and the time of starting was mainly caused by financial and other arrangements managed by Uville. On the 1st September, 1814, Uville, Henry Vivian, Thomas Trevarthen, and William Bull sailed from Portsmouth for Lima in the ' Wildman,' taking with them four pumping engines, with pump - work and rods complete; four winding whim-engines, with all winding apparatus complete; one portable locomotive engine on wheels, to be used for a rolling mill or other purposes; one mill for grinding ore; and one rolling mill, probably for the Mint at Lima. These nine steam-engines, with their apparatus complete for work at the mines, cost 6838*l.*; the grinding and rolling mill cost 700*l.* more; but various other expenses more than doubled the amount, which reached the large sum of over 16,000*l.*

On reference to the conditions of agreement under which Uville acted, dated 17th July, 1812, Don Pedro Abadia, Don José Arismendi, and Don Francisco Uville, were partners engaging to drain a range of mines. Uville was to go to London to purchase two steam-engines, and was authorized to expend $30,000 (say 6000*l.*). $2000 (say 400*l.*) was to be paid to him as the value of Trevithick's model, which he had a few years before bought in London for 21*l.* He was to engage one or two English workmen. No new partner was to be allowed. They also contracted with the various workers of mines in Yauricocha, Yanacancha, Caya Chica, Santa Rosa, and in the mining ridge of Colquijilca, for a period of nine years, to commence within eighteen months of that time, to sink a general pit for the drainage of those mines, and to pump out the water by steam-engines. The payment for this drainage was to be one-twentieth part of the ore raised by the different mines.

" An agreement made at London this 8th day of January, 1814, between Don Francisco Uville, of Lima, in the Vice-royalty of Peru, of the one part, and Richard Trevithick, of Camborne, in Cornwall, engineer, of the other part. Whereas, by an agreement of partnership made and signed at Lima, and whereas the said Francisco Uville did in pursuance of his con-tract with the said miners soon after the ratification thereof, embark for England, for the purpose of fulfilling the same on his part, and on his arrival there in the month of April last, made application to the said Richard Trevithick, who is an experienced engineer and miner, and requested him to assist him in promoting the object of his journey, which the said Richard Trevithick (being penetrated with a high sense of its utility) agreed to do, and hath accordingly applied himself wholly to that object, ever since the arrival of the said Fran-cisco Uville in England: And whereas under the direction of the said Richard Trevithick, and by the orders of the said Francisco Uville, various machines and engines have been made for the purposes of the said concern, a part of which has been already paid for by the said Francisco Uville; but several of the bills brought by him to England not having been honoured, by reason of the absence from England of the parties upon whom they were drawn, the said Francisco Uville hath not at present sufficient funds to answer the engagements he has entered into in this country, and Don Juan . . . . , to whom he was in that case directed by his partners to offer shares in the said concern, and from whom he could have received supplies, not being at this time in London, the said Francisco Uville has agreed to admit the said Richard Trevithick to be a partner in the concern, upon his advancing and paying a proportionable part of the expenses necessary for carrying on the same. Now these presents witness that in consideration of the said Richard Trevithick having paid and agreeing by these presents to pay certain bills for machinery ordered by the said Francisco Uville to the amount of 3000*l*. or thereabouts, the particulars of which have been ascertained and settled by and between the said Francisco Uville and Richard Trevithick, and also in consideration of the services which the said Richard Trevithick hath already rendered to the said undertaking, and of the future benefits which he is expected to perform for it, the said Francisco Uville for himself, and on the behalf and in the name of the said Pedro Abadia and José Arismendi (who will ratify these presents in the capital of Lima as soon as it shall be produced to them, to which the said Uville holds himself bound), Doth, by virtue of the power and authority given to him by his said partners, agree to admit the said Richard Trevithick to be a member of the said company, and doth hereby declare him to be a member thereof and a partner therein to the extent of 12,000 dollars, and as such, entitled to a share and interest in all the profits and advantages of the company in the proportion which the said sum of 12,000 dollars shall bear to the amount of capital employed by the company in the purposes of their establishment, which proportion will amount as nearly as can now be ascertained to one-fifth of the capital stock embarked in the said concern.

"FRAN. UVILLE.

"RICHARD TREVITHICK.

"8th *January*, 1814."

So Trevithick paid 3000*l*. and received nothing for his engineer's work, to be made a partner, contrary to Uville's limit of authority, in a speculation that proved to be not worth a farthing.

The following is a summary of the detail invoice of engines and machinery which left London for Lima in September, 1814, in charge of Uville, just fifteen months after his landing at Falmouth in search of Trevithick:—

"Invoice of four steam-engines, four winding engines, one portable rolling engine and materials for ditto, two crushing mills, four extra-patent boilers, spare materials for engines, boring rods, miners', blacksmiths', and carpenters' tools, &c., shipped on board the 'Wildman,' John Leith, master, from London to Lima, by, on account and risque of Don Francisco Uville, Don Pedro Abadia, and Don José Arismendi, merchants at Lima.   Dated 1814.

		£	s.	d.
To four steam-engines of 33-horse-power each (complete for lifting water with under-adit and house lift-pumps, and wrought-iron pit-work, rods, &c., at 1399*l.* 13*s.* each ..		5,598	12	0
To four winding engines of 8-horse-power each, with whims, barrels, shafts, &c., complete for lifting ore, at 210*l.* each		840	0	0
To one portable steam-engine of 8-horse power, for rolling, with its chimney, axles, carriage-wheels, &c.		400	0	0
		6,838	12	0
A mill for grinding ore	£517 0 0			
A rolling mill	204 0 0			
Duplicates, sundries, freight, insurance, &c., &c.	8,592 9 1			
		9,313	9	1
		£16,152	1	1 "

The nine steam-engines, including a locomotive, with its chimney, axles, carriage-wheels, &c., a crushing mill and a rolling mill, cost but 7560*l.*   Other expenses, for freight, insurance, &c., &c., increased the amount to 16,152*l.*

William Williams,[1] on his return from the Cerro de Pasco Mines, states :—

"On the 3rd March, 1872,.I saw in Yauricocha Mine two of Mr. Trevithick's engines at work; one of them was a horizontal 12-inch open-top cylinder pumping engine, about a 4-feet stroke; there were two fly-wheels about 10 feet diameter and a cog-wheel 7 feet diameter, giving motion to two wrought-iron beams working a 10-inch pump bucket.   The other was a 12-inch cylinder winding engine with a large fly-wheel.   Three Cornish boilers, about 5 feet 6 inches diameter, with 3 feet 9 inch tube, 30 feet long, made of $\frac{7}{16}$ths of an inch plates, supplied steam of 40 lbs. on the inch."

---

[1] Residing at Angarrack, near Hayle, 1872.

CHAPTER XXII.

PERU.

" CONDITIONS under which Don Pedro Abadia, Don José Arismendi, and Don Francisco Uville, establish the project of draining the mines by means of steam-engines, to be brought from England.

" 1st. The company is composed of three contracting persons without admitting therein any other whatever.

" 2nd. There are intended as a fund for the undertaking 40,000 dollars, to be divided into four shares in the following manner :—Two shares to Don Pedro Abadia, one to Don José Arismendi, one to Don Francisco Uville. Four shares, dollars 40,000.

" 5th. These principles of good faith and friendship being established, the project is to be carried into effect with the greatest possible activity, for which purpose, by the first opportunity, the funds shall be forwarded by Don Pedro Abadia to the amount of 30,000 dollars, with the necessary instructions for the construction of the machinery to a person who may be appointed.

" 7th. As it has been estimated that 30,000 dollars will cover the cost of two engines in England, if the said Uville finds another on credit, he is authorized to purchase it on account of the company.

" 11th. Should the undertaking yield profits, Uville shall also be credited for 2000 dollars for the value of the model.

" 12th. In the instructions that may be given to Uville, it shall be stipulated on what terms he may engage one or two English workmen.

" LIMA, 17th July, 1812."

" *Contract.*

" 1st. The present contract shall be considered binding for nine years, to be computed from the time the steam-engines may be erected in the different parts of these mines that may be judged suitable.

" 2nd. The miners herein contracting cede their mines in Yauricocha, Yanacancha, Caya Chica, Santa Rosa, and in the mining ridge of Colquijilca, and the company offer the means, steam-engines, and instruments for draining the same, and on these principles the obligations of both parties are as follow, to wit.

" 3rd. The company binds itself within the period of eighteen months, or sooner if possible, to bring over the steam-engines to drain successively the different parts of these mines, and immediately on their arrival to place them in Yauricocha, and afterwards in Yanacancha, Caya Chica, Santa Rosa, and in the mining ridge of Colquijilca, to sink a general pit for the collection of the waters at a depth of 40 varas from the adit or drainage level of Santa Rosa.

" 8th. Each miner whose mine situated in the parts above specified is not perfectly drained in consequence of the filtration or natural gravity of the water to the general pit, is to continue a tube to communicate with the said general pit on his own account, in order fully to enjoy the benefit of the draining, it being well understood that the company shall not refuse to admit the waters of any of the mines situated in this part whatever their quantity may be. And the company shall be further bound to supply funds to any miner who may not have sufficient to defray the expenses of such tube of communication at an interest of 6 per cent., to be refunded out of the first metals which may be obtained.

" 10th. The recompense to be made to the company for the general drain procured in the place or places agreed on, shall be, with regard to Yanacancha and Yauricocha, in consequence of the known richness of those places, and of the timber required by the softness of the ground to secure the mines, 15 per cent. on the ore that shall be extracted therefrom, and lodged either in the common depôts or in the respective warehouses; and in the mines of Santa Rosa, Caya Chica, and Colquijilca, 20 per cent., which distribution is respectively to be made on the quantities obtained.

"14th. That the miner who refuses to enter into this fair contract whose mines are benefited by the means of the engines, shall be compelled to pay the contributions and to perform what has been therein stipulated according to ordinance.

"This contract being agreed to, the contracting parties signed respectively to be bound and compelled; and I, the Royal Judge and Sub-delegate hereof for His Majesty, signing it with all the contracting parties and witnesses before me on the said day, month, and year.

"Pedro Abadia, José Arismendi, Francisco Uville, José Maria de Ulloa, Ignacio Beistequi, The Marquis de la Real Confianza, José Herresæ, Publo Anellfuertes, Ramon Garcia de Purga, José Antonio de Arrieta, José Camilo de Mier, José Lago y Lemus.

"For myself and Don Remiqia, p. procuration Manuel Queypo, Rafael Doper, Juan Gonzalez, Augustin Zambrano, Francisco Rasines, Francisco Fuyre, Manuel Ysasi, Alberto de Abellaneda, Ysidro Crespo, Juan Antonio Arrasas, Pedro Gusman, Manuel Yglesias, Patricio Bermudez, Bartolome de Estrada. For the miners, Don Castano Villanueva, Juan Isidoro, Manuel de Santalla, Juan Palencia, Antonio Perez, Manuel Cavellero, Domingo Pallacios, Matias Canallero, Ambrosio Ortega, Francisco de Otayequi, Pedro de Arrieta, Juan de Erquiaga, José Zeferino Abaytad, Antonio Villaseca, Estanislas Maria de Arriola, José Maria del Veto, Ambrosio Guidones, Santiago Oreguela. For Don Pedro Mirales, p. procuration, Thomas Hidalgo, Nicholas Berrotarran, Barnabe Perez de Ybarrela, Augustin Bayroa, Francisco Xavier de Uribe, Manuel Varela. For my brother, Juan Francisco de Aspiroz, Juan Miguel de Aspiroz.

"In the city of Los Reyes on the 26th September, 1812."

These extracts from an agreement drawn up by the leading men in Peru in 1812 are proofs of remarkable energy. Rumours of the power of steam-engines used in mines in England had reached Lima, Don Francisco

Uville was sent on a mission of inquiry, and in 1811 consulted Boulton and Watt at Soho, who gave an opinion that their engines were not suitable to so elevated a position where the atmosphere was so much lighter than in England, and the difficulties of transit so great. On his return to Lima he carried with him a small model of Trevithick's high-pressure steam-engine. The Spaniards on seeing it work had the good sense and courage to put aside the Watt report and adopt the principle of the small but active high-pressure steam-puffer engine.

An influential company was formed, which sent Uville again to England to seek out the high-pressure engineer and purchase his engines. What stronger evidence could be given of the great difference between the rival engineers and their engines? The one with low-pressure steam and vacuum, the other with high-pressure steam and without vacuum.

The three persons contracting to drain the Peruvian mines agreed that no other should be allowed to join them in the contract; two steam-engines were to be purchased, and if convenient a third engine might be ordered on credit. One or two English mechanics were to accompany the engines which the contractors engaged should be in Lima within eighteen months. Ten months had passed before Uville reached Trevithick, and when in May, 1813, he communicated to the Cornish engineer the same wants that he had made known to Watt two years before, how different was the answer received. "I engage to supply in four months six 24-inch cylinder high-pressure steam pumping engines, with pumps and all necessary apparatus complete."[1]

[1] See Trevithick's letter, 22nd May, 1813, vol. ii., p. 198.

This promise was nearly fulfilled,[1] but want of money, the ordering of additional machinery, and difficulty in finding a ship,—for Spain was then at war, or on the verge of it, with the South American republics,—delayed for a time the completion of the order; but within eight months even the additional work seems to have been ready, and the following agreement was entered into, though the ship with her freight of *nine* steam-engines did not leave England until September, 1814, fifteen months after Uville's first meeting with Tre-vithick.

"*Agreement dated the 8th January*, 1814.

" The said persons from whom he (Uville) would have received supplies, not being at that time in London, the said Francisco Uville has agreed to admit the said Richard Trevithick to be a partner in the concern, upon his advancing and paying a proportionable part of the expenses necessary for carrying on the same. Now these presents witness, that in consideration of the said Richard Trevithick having paid, and agreeing by those instruments to pay certain bills for machinery ordered by the said Francisco Uville to the amount of 3000*l.*, and also in consideration of the services which the said Richard Trevithick hath already rendered, and of the future benefits which he is expected to perform, doth agree to admit the said Richard Trevithick a partner therein, as nearly as can be ascertained to one-fifth share of the whole.

" He hath planned and directed the particular construction of three steam-engines, and hath for that purpose taken many journeys to manufacturing towns and other places.

" He hath given to the said Francisco Uville a general know-ledge of English mining, miners' tools, winding and crushing engines, &c., &c., and for that purpose hath taken him to various mines in England, to which the said Richard Trevithick, through his interest, had access. He hath instructed the said Francisco

[1] See Trevithick's letters, 22nd Sept. and 23rd Oct., 1813, vol. ii., pp. 206, 209.

Uville in the art of making drawings of mines, and in engineering.

" He hath furnished him with various drawings of English mines, and plans for the future working of Spanish mines, and hath given to him every other engineering and mining information.

" He hath increased the power of the three engines above mentioned to the extent of one full third, without making any additional charge for so doing, and he hath agreed to supply the said company with a fourth engine, and to wait for the payment of it, until the return of the said Francisco Uville to Lima, in recompense for all which the said Francisco Uville doth for himself and his partners grant to the said Richard Trevithick one and quarter per cent. of the net produce or profits (all expenses first deducted) of the ore extracted from the said mines, and as a further recompense, doth appoint him sole engineer in Europe for all the machinery that shall be used or required."

The nine steam-engines, with apparatus for minting, crushing ores, draining, winding, and even locomotion, with miners' tools complete down to mine ladders, borers, picks and gads, and hammers, were received by a large and influential body of Spaniards residing near Lima, under the special patronage of the Viceroy. The machinery had then to be taken up precipitous tracks that foot-passengers trembled to walk on, to the height of more than 15,000 feet.

The calculated profit was 500,000*l.* a year, of which 100,000*l.* a year was to be Trevithick's share, a portion of which was sold to pay for the engines. A prospectus drawn up in England states that " the whole capital was in four hundred shares, of which Trevithick held eighty, valued at 40,000*l.*, together with special advantages to be accorded to him."

The machinery having left England in September,

1814, reached Peru in the early part of 1815, shortly after which one of the engines was at work in the Mint at Lima, within two years from the giving the order for it in England; for in the early part of the latter year Trevithick wrote to one of his men :—

"I am sorry to find by Mr. Uville's letter that the Mint engine does not go well. I wish you had put the fire under the boiler and through the tube, as I desired you to do, in the usual way of the old long boilers, then you might have made your fire-place as large as you pleased, which would have answered the purpose, and have worked with wood as well as with coal, and have answered every expectation.

"I always told you that the fire-place *in the boiler* was large enough for coal, but not for wood, and desired you to put it under it. The boiler is strong enough and large enough to work the engine thirty strokes per minute, with 30 lbs. of steam to the inch. I hope to leave Cornwall for Lima about the end of this month, and go by way of Buenos Ayres, and cross over the continent of South America, because I cannot get a passage; none of the South Sea whalers will engage to take me to Lima, they say that they may touch at Lima or they may not, in the whole course of their voyage; therefore, unless I give them an immense sum of money for my passage, they will not engage to put me on shore at Lima, and for me to risk a passage in that way, and to be brought back again to England after two years' voyage, without seeing Lima, would be a very foolish trip; therefore to make a certainty, I shall take the first ship for Buenos Ayres, preparations for which I have already made." [1]

The whole of the machinery having been sent off, Trevithick was prepared to make his way across the then little-known continent of South America in its broadest part, from Buenos Ayres to Cerro de Pasco. [2]

---

[1] Unfinished rough draft of letter by Trevithick.
[2] See Trevithick's letter, December 9th, 1815, vol. ii., p. 31.

His departure was deferred from various causes until the 20th October, 1816, when he sailed from Penzance in the South Sea whaler 'Asp,' Capt. Kenny.

PENZANCE IN OLDEN TIME. [W. J. Welch.]

"DEAR SIR,   "PENZANCE, 20th August, 1817.

"I am enabled to furnish you with a few particulars which led to the introduction of steam-engines into Spanish America, which you will embody into your interesting paper for our next Geological meeting, as you deem most proper.

"Captain Trevithick was born in Illogan, Cornwall, 1771, but he has generally resided at Camborne, the adjoining parish. He has devoted the greatest part of his life to mechanics and to improvements in the high-pressure steam-engine, and many engines of Captain Trevithick's construction are now working in different parts of England.

"Mr. Francisco Uville, a native of Switzerland, visited Lima and the rich Peruvian mines in the neighbourhood of Lima, at an early age, and being a gentleman of great intelligence, he thought it possible that the silver mines at Pasco, about 150 miles from Lima, which were fast falling into decay for want of machinery to drain the water, might be restored to their former celebrity by the introduction of steam-engines.

"Mr. Uville, who is now about thirty-six years of age, came to England in 1811, where he continued a few months, and just as he was about to leave London he observed by accident a model of a steam-engine, made by Captain Trevithick, at the shop of a Mr. Roland, Fitzroy Square, and Mr. Uville so much liked the simplicity of its construction, that he immediately purchased it at twenty guineas. Mr. Uville returned to Lima with it, and tried it on the mountains of Pasco, in consequence of which, on the 17th of July, 1812, Mr. Uville, with Don Pedro Abadia and Don José Aresmendi, eminent merchants at Lima, were so confident of success, that they formed a company to drain the mines at Pasco and its vicinity; and on the 22nd of August then following a contract was entered into by these gentlemen and the proprietors of the mines in that district. Soon after which Mr. Uville was deputed by the company to return to England and to find out some able engineer to assist him in procuring proper steam-engines to be conveyed to the mines.

"Uville having put into Jamaica, came to England in the 'Fox' packet, Capt. Tilly, and arrived at Falmouth early in the summer of 1813. During the passage Mr. Uville frequently talked of the object of his voyage, and that he was particularly anxious to find out the maker of the model of the engine he took to Lima, and recollecting that the name of 'Trevithick' was on the model, he mentioned it to a Mr. Teague, who happened to be on board the packet, when the latter informed him that Capt. Trevithick was his first cousin, and that he resided within a few miles from Falmouth. Immediately on Mr. Uville's arrival an interview took place between him and Capt. Trevithick, and soon after Mr. Uville removed to Capt. Trevithick's house in Camborne, where he resided several months, during which time Capt. Trevithick instructed him in mining, machinery, &c.

"Capt. Trevithick and Mr. Uville, after seeing most of the

mines in Cornwall, visited several other mining districts in
England, to afford Mr. Uville a better opportunity of acquiring
the best knowledge of engineering by examining the steam-
engines erected. Afterwards they went to London, when Mr.
Uville was introduced to a Mr. Campbell, of the East India
Company's department. Mr. Campbell informed Mr. Uville
that the best engineers in Europe were Messrs. Boulton and
Watt, of Birmingham; and strongly recommending them to
him, he observed that he was convinced if engines could be
made capable of being transported to the mines of Pasco across
the mountains they would be able to do it. Mr. Uville accordingly applied to these gentlemen, and fully explained to them
the nature of the engines which would be wanted, and the state
of the road by which they must be conveyed, and Messrs. Boulton
and Watt returned an answer that it would be impossible to
make engines small enough to be carried across the Cordillera
to the mines.

"Capt. Trevithick, however, was not startled at the difficulties, and having applied himself to the improvements of his
high-pressure engines, entered into a contract with Mr. Uville
to provide nine steam-engines for the company at Lima; and,
by virtue of the powers with which Mr. Uville was invested,
Capt. Trevithick was admitted a partner of one-fifth in the
concern; besides which, for his great pains and services he had
rendered, Mr. Uville guaranteed to him a handsome percentage
on the profits of the company (*vide* Articles of Agreement of
8th January, 1814).

"These matters being settled, nine engines were provided at
an expense of about 10,000*l*., and were shipped on board the
'Wildman,' South Sea whaler, Capt. Leith, who sailed from
Portsmouth for Lima the 1st September, 1814, accompanied
by Mr. Uville and the following Cornish engineers,— Thomas
Trevarthen, of Crowan; Henry Vivian, of Camborne; and William Bull, of Chacewater, in Gwennap.

"The engines arrived at Lima, and were received by a salute
from the Government batteries, and the greatest joy was testi-
fied on the occasion.

"On the 27th July, 1816, the first steam-engine was set to

work at Santa Rosa, one of the mines of Pasco, under the direction of Mr. Bull (*vide* despatch of that date, signed José G. de Prada).

"On the 20th October, 1816, Capt. Trevithick sailed for Lima in the 'Asp,' South Sea whaler, Capt. Kenny, accompanied by Mr. Page, a gentleman of London, and James Saunders, of Camborne, an engine maker; and on the 6th February, 1817, they arrived at Lima, where Capt. Trevithick was immediately introduced to the Viceroy by Don P. Abadia, and he received the most marked attention from the inhabitants (*vide* 'Lima Gazette' of 12th February).

"Perhaps you will think it proper to notice the furnaces which Captain Trevithick took out in the 'Asp' to Lima for the purpose of purifying the silver by sulphur. A great expense will be saved by these means. Any further information which I can afford you I will readily give.

"I am, dear Sir,

"Your very obedient and humble servant,

"RD. EDMONDS.

"H. F. BOAZE, Esq."

This statement, from a solicitor more than fifty years ago, inadvertently points out the difference between the steam-engine of Watt and that of Trevithick. The former said it was impossible to make engines having the required power small enough to be carried to the mountain mines, whereas a small high-pressure engine by the latter had sufficient power.

Day and Page were lawyers advising Mr. Uville in London. Page sailed from Penzance with Trevithick and James Saunders, a boiler maker, in the 'Asp,' a South Sea whaler, on the 20th October, 1816, just two years after the departure of Uville with the machinery and engines. The difficulty of conveying heavy weights up the mountain foot-paths was almost insurmountable.

Mr. Rowe, who went to these mines in 1850, says,—

"The Cerro de Pasco mines are about 170 miles from Lima; we crossed a ridge 25,000 feet high. The mines were about 13,400 feet high above the sea. There was but one road; no wheel vehicle could be used; 'everything was carried on mules. Sometimes the road was only $2\frac{1}{2}$ feet wide, cut in precipices three or four hundred feet perpendicular: some of the men were afraid to walk, and dared not ride.

"I lived in the house that used to be Mr. Trevithick's office and store-room; it was in the suburbs of the town of Cerro de Pasco. The shafts are some of them in the middle of the town; several pieces of Captain Trevithick's engines lay about the shafts, and some on the way up, as though they had stuck fast, and some we saw at Lima. Mr. Jump, a director on the mine, pointed out a balance-beam that Mr. Trevithick had put up thirty years before. Only one Englishman then remained there who had worked for Mr. Trevithick; he was called Sycombe, and said Trevithick's men were an unmanageable lot.

"The natives worked in the mines underground. The atmosphere was only about 10 lbs. on the inch. We found a coal mine not far off; the quality was not very good. The smiths had difficulty in welding with it. Our heaviest pieces of machinery did not exceed 280 lbs. The worst parts of the road have been a little improved since that time."

Just one month before Trevithick sailed from Penzance for Lima, the first pumping engine taken out by Uville had been satisfactorily put to work in the mountain mine of Santa Rosa, with its steam-cylinder weighing double the limit fixed on by modern engineers.

The following information respecting the progress of the steam-engine fixed on the Santa Rosa Mines, one of the mineral ridges of Pasco, in the Viceroyalty of Peru, is extracted from the Government Gazettes of

Lima, dated the 10th of August and 25th of September, 1816 :—

"PROGRESS OF THE STEAM-ENGINE, &c.

"*His Excellency the Viceroy of Peru to the Editor.*

"In order to satisfy the eager expectations of the inhabitants of this Viceroyalty, those of the greater part of these Americas, and even of the Peninsula itself, I hereby order the printing, at full length, in the next Government Gazette, or at same time in a separate sheet, the enclosed despatch from the Intendant Governor of Tarma, giving the details of the admirable results of the steam-engine fixed in the mineral territory of Pasco, for the most important purpose of draining its mines, and for the extraction of its rich ores. This authentic communication must produce the most lively and grateful sensations in those true Spaniards, who with grief contemplated as irreparably lost the only spring from which flowed the prosperity of this continent, excite their just acknowledgments to the meritorious co-operators in such an expensive and difficult as well as eminently-advantageous enterprise, and encourage to similar undertakings in other parts those who, with personal aptitudes and patriotic sentiments, have been waiting the final success of the first.

"JOAQUIN DE LA PEZUELA.

"LIMA, *4th August*, 1816."

"*Certificate of the Deputation.*

"We, Don Domingo Gonzales de Castañeda and Don José Lago y Lemus, Commissaries and Territorial Magistrates in this Royal Mineral Territory, and deputed by the United Corporation of Miners in this district, do hereby certify judicially, and as the law directs, in manner following :—

"Though this deputation never doubted the extraordinary power of steam compressed, and consequently the certain operation of engines worked by its influence, it nevertheless entertained some fears respecting the perfect organization of all the mechanical powers of the machines. This uncertainty, rather than any doubt, has been completely dissipated by our personal attendance this day to witness the draining of the first pit, situated in Santa Rosa. The few instants employed in the same produce a full conviction that a general drainage of the mines will take place, and that their metals will be extracted with the greatest facility from their utmost profundity: as also that the skill of the company's partners and agents will easily overcome whatsoever difficulties nature may oppose, until they shall have completed all the perpendiculars and levels; and consequently that the meritorious undertakers who have risked their property in the enterprise will be rewarded with riches.

"We and the whole Corporation of Miners would do but little were we to erect them a monument, which should transmit down to the remotest posterity the remembrance of an undertaking of such magnitude and heroism; but for the present we will congratulate ourselves that our labours, co-operation, and fidelity, keeping pace in perfect harmony with the exertions of the agents, the company may thus attain the full completion of their utmost wishes, extracting from the bowels of these prolific mountains, not the riches of Amilcar's inexhaustible wells, not the treasures of the boasted Potosi in its happiest days, but a torrent of silver, which will fill all surrounding nations with admiration, will give energy to commerce, prosperity to this Viceroyalty and to the Peninsula, and fill the royal treasury of our beloved sovereign.

"Thus certifies this Magisterial Deputation of Yauricocha, the 27th of July, 1816.

"DOMINGO GONZALES DE CASTANEDA.
"JOSÉ LAGO Y LEMUS."

"*Despatch from the Intendant Governor of the Province of Tarma to His Excellency the Viceroy.*

"MOST EXCELLENT SIR,　　　　　　　"PASCO, 27*th July*, 1816.

"Having finally conquered the great difficulties consequent on the enterprise, though with immense and incessant labour, and at an enormous expense, the object has been accomplished of purchasing, importing, and erecting the steam-engine in the celebrated rich and royal mineral territory, called ' The Mountains of Yauricocha,' in this province of Tarma, of which I have the honour to be Governor, the chief and valuable works of which have ceased to produce ore, in consequence of their bases being completely submerged in water.

" The day is arrived when we witness with admiration the advantageous and useful effects of the before-named steam-engine; the completion of the promises made by the generous and undaunted individuals who united themselves to supply the funds sufficient for the realization of an enterprise so important, and the fulfilment of the wishes of these valuable subjects, to render to the State the highest possible service; a service, although at all times of extreme importance, at this crisis is infinite; because the State, being weakened by a series of disastrous events for six years past, requires salutary remedies; and none exist so effectual as the re-establishment of the mines, which the steam-engines are achieving.

" After some experiments, which (although they left no doubt of ultimate success in draining the mine) discovered some slight defects, these were corrected on the 23rd instant; and this day the first of the four pumps which arrived for the use of the royal mines was erected in the particular mine called Santa Rosa; the result of its operation has been the exhaustion of the water from the well or hollow below the adit. In twenty minutes, by this engine, an aggregate of water is ejected amounting to 6 yards or 18 feet in diameter, 3 yards 24 inches in length, and 1 yard 30 inches in breadth. In the same manner a second engine, accessory to that which drains the water, is worked by the same steam, on the same point, and in the same perpendicular shaft, from the surface of the earth, which extracts the ore, and with advantages hitherto unknown here, on account of the considerable saving of expense and the economy of manual labour.

" The steam-engine will continue evacuating the water from the pit until it is reduced to 6 yards below the old adit, whence they must eject the water raised by the engine by a continued elongation of the barrel of the pump gaining depth, until they have completed the number of yards required, or until the progress of the work indicates a proper situation for forming a new line of levels and channels of communication to those mines which are not yet drained. In proportion to the successive acquisition to these subterranean works which are daily advancing, will be the increased operations of the mines, and consequently the increased prosperity of the mining interest, which had most astonishingly fallen from the degree it had attained in former years.

"In a short time, similar effects will be seen in the three remaining mineral ridges of Yauricocha, Caya, and Yanacancha, productive of ores of a better quality than that of Santa Rosa, which has nevertheless obtained the preference for the erection of the first engine upon it from its being more abundant in its peculiar produce, and on account of the greater number of persons interested in this property; as also its contributing immediately to relieve the necessitous, by employing the workmen in the vicinity of the mines. In my opinion, no event so beneficial has occurred as the erection of the steam-engine, since the discovery and addition of these dominions to the crown of Castile. From this time, by the help of these machines, immense and incalculable riches will accrue to the nation.

"God preserve your Excellency many years,

"José Gonzales de Prada."

"*The Viceroy's Answer.*

"Lima, 5th August, 1816.

"Your Lordship's official despatch, No. 898, the 27th of last month, communicated to me the satisfactory detail of the complete results which you witnessed on the 23rd and 27th, produced by the grand steam-engine placed over the mine of Santa Rosa, one of those situated in the mountains of Pasco, for the purpose of draining the mine and extracting the ores.

"I desire particularly to distinguish and patronize the chief agent and assistant, Don Francisco Uville; the generous promoters of the undertaking, Don Pedro Abadia and Don José Arismendi; their agents and assistants, Don Luis de Landavere and Don Tomas Gallegos; and lastly, Mr. Bull, and all those associated in this great work, whom you recommend to my notice. Your Lordship, by having exerted yourself to facilitate, by all the means which your zeal and authority could procure, the happy consummation of so profitable an enterprise, has added a new claim to the many preceding which you possessed, to the high consideration of the king and the public.

" JOAQUIN DE LA PEZUELA,
" Viceroy of Lima."

Extract from the ' Lima Gazette' of the 25th of September, 1816.

" We have the satisfaction of communicating to the public the information that the company for draining the mines of Pasco have just received accounts from their agents in that mineral territory; and they promise for our next Gazette a description of the state of the works for fixing the remaining three engines.—EDITOR.

" CERRO, *September 20th,* 1816.

" ' After having observed the progress of the machine at the Santa Rosa Mine last Saturday, the 14th instant, at 10 o'clock at night we began to act; at 11 o'clock the pitmen went down to clear the shaft, and have not since ceased working an instant. The clearing of the mud and rubbish which had remained at the bottom of the shaft, and clogged every moment the buckets and suckers of the engine, lasted till Wednesday; but this being accomplished, at 12 o'clock at noon they began to break through the level. At half a yard below the shaft we found a lively coppery ore, with its particles of silver. This bronze-coloured ore indicates that the veins of Yauricocha and San Diego incline to the west, or towards the Santa Rosa Mine. The mines in the vicinity of this pit are all dry. Some of them, at the distance of 300 yards, in the ridge of Santa Rita, have also felt its effects; and even as far as the territory of Caya, behind our steam-works, the waters have fallen in several mines. Don John Vivas has begun to work in San Diego Mine. They are also going next Monday to begin working in several points of the Santa Rosa Mine. The pit is already 8 yards in depth, and we are proceeding with the greatest activity. The workmen are relieved every two hours, and as they go out they give up their tools to those who succeed them, by which means not a minute is lost. Continuing thus, in the course of a month we shall be at more than 20 yards depth, and have many mines in full activity. The winding engine raises a basket (which is a load) in two minutes; the draining or steam-engine, with two vibrations per minute, keeps the surface always dry. Both work with the greatest ease, certainty, and regularity.

"'By dint of searching after a vein of coal, we have at last found one near at hand, of excellent quality and of great richness. The pit we are now at work at is at the distance of a quarter of a league from Rancas, and at the same distance from Vista Alegre which the Cerro is from these works. We have likewise found a vein of plumbago, which was an object of search, on the supposition that it was coal. This substance, of which much is consumed, mixed with grease, to soften the friction of the piston, &c., we have now here; and thus the necessity of sending to Lima, or perhaps to Europe, for it is obviated.'"

Within six months of the setting to work the pumping engine in Santa Rosa, another pumping engine was at work at Yanacancha Mine.

The following extracts from the 'Lima Gazette' were published in the Cornish papers by Mr. Edmonds:—

"*From the Government Gazette of Lima*, 12*th Feb.*, 1817.

"We have had the pleasure of receiving a letter from Pasco, dated 6th instant, containing the following account:—

"The second engine established in the mine called Yanacancha, which is far superior in point of beauty, convenience, and size to that called Santa Rosa, was set to work on Friday last, and notwithstanding the great quantity of water which filtered into this mine the engine with only half its power drained the mine completely in nine minutes. This filtration did not happen in Santa Rosa, on account of the quantity of hard copper ore on which the engine is situated.

" By this successful operation, the water in several mines has been lessened considerably, amongst which in particular is that belonging to Don Juan Vivas, situate in the hill called Chuca-rillo, which at present affords ore of 400 marcos per caxon (50 cwt.). Of this ore about 25 lbs. has been received in this city, with a proof of 2 lbs. made in Pasco, showing not only the richness of the ore, but its easy extraction and cleanness for the ready refinement of it. And another proof has also been received from another mine, situate in Chucarillo, belonging to the widow Mier, in company with Don Joachin Aitola, which yields 100 marcos per caxon of 25 cargas.

" To this agreeable news we ought to add that at the arrival of the whaling ship ' Asp,' bound from London, having on board a large quantity of machinery for the Royal Mint, and for the constructing of eight engines more, equal to those in Pasco, with the advantage that they are of the last patent and more easy to be worked; but what is of greater importance is the arrival of Don Ricardo Trevithick, an eminent professor of mechanics, the same who directed in England the execution of the machinery now existing in Pasco. This professor can, with the assistance of the workmen who accompany him, construct as many engines as are necessary in Peru, without any need of sending to England for any part of these vast machines. The excellent character of Don R. Trevithick, and his ardent desire for promoting the interests of Peru, recommend him in the highest degree to public estimation, and make us hope that his arrival in this kingdom will form the epoch of its pros-perity, with the enjoyment of the riches enclosed in it, which could not be enjoyed without this class of assistance, or if the British Government had not permitted the exportation from England, which appeared doubtful to all those who knew how jealous that nation is in the exclusive possession of all superior inventions in arts or industry."

So far everything promised success. Two pumping engines had so reduced the water in two of the mines, that the miners were at work, and the people of Lima believed that many more such engines would be usefully

employed, now that Don Ricardo Trevithick was with
them.

"DEAR SIR,   "LIMA, *February 15th*, 1817.

      "We arrived here last Saturday in good health. The
(our) Mint is at work, and coined five millions last year, and
in their way of working does very well; but I trust to make
it coin thirty millions per year.

      "Two engines are drawing water, and two drawing ore, at the
mines, but in an imperfect state. If I had not arrived, it must
have all fallen to the ground, both in their mining and in their
engines. I expect we shall go to the mines in about ten days,
from where I will write to you every particular.

      "There are still two engines to put up for lifting water, and
two for winding ore, and those at work to be put to rights. They
are raising ores from one mine which is immensely rich, and
from what I can learn, a much greater quantity will be got up,
when the whole are at work, than these people have any idea
of. Several other mines will also be set to work by engines
that we shall make here. We have been received with every
mark of respect, and both Government and the public are in
high spirits on account of our arrival, from which they expect
much good to result.

      "Mr. Vivian died the 19th of May. I believe that too much
drink was the cause of it. Uville, I think, wished him gone,
and was in great hope that I should not arrive. His conduct
has thrown down his power very much, which he never can
again recover.

      "They all say that the whole concern shall be put entirely
under my management, and every obstacle shall be removed out
of my road. Unless this is done, I shall soon be with you in
England. I am very sorry that I did not embark with the first
cargo, which would have made a million difference to the company. The first engine was put to work about three months
since, the other about two months; but they are as much at a
loss in their mining as in their engineering. The Mint is the
property of our company, and Government pays us for coining,
which gives us an immense income; the particulars of which,

and the shares in the mines, I have not yet gone into. I shall be short in this letter, because I know but little as yet, and that little I expect Mr. Page will inform you. A full account you shall have by the next ship, which I expect will sail in three weeks. This letter goes by a Spanish ship that will sail this afternoon for Cadiz. My respects, and good wishes to your family and to Mr. Day, and hope this will find you all as hearty as we are.

"Mr. Page would not depart this life under the line, as he promised when at Penzance; but, on the contrary, has a nose as red as a cherry, and his face very little short of it. His health and spirits far exceed what they were in England. I am glad to have such a companion. With . . . . think he will have no reason to repent . . . . He will get a command at Pasco . . . . such as his ingenuity may find out, when on the spot; whether as a miner or an engineer I cannot say, but time will show.

"If you have not insured my life I would thank you to do it now, if you can on reasonable terms. I do not wish them to take the risk of the seas in the policy, because the voyage here is over, and on my return I hope I shall not want it, therefore it must be for two years in the country. I will get a certificate of my health, if they wish it, from the most respectable inhabitants, and also from the Vice-king, if they wish it. The policy may be drawn accordingly.

"Be so good as to write me often, with all the news you can collect. If you wish your dividends in this company to be applied to further advantage in any new mines I may engage in, in preference to having it sent to England, I will, as the dividends are made, do everything in my power to improve the talent. On this subject I must have your answer before I can make any new arrangement under this head. I will thank you to send a copy of Mr. Page's letter to my wife; I mean such parts of it as belong to the business; there may be some things that I have forgotten to mention.

"I remain, Sir,

"Your humble servant,

"RICHD. TREVITHICK.

"MR. JAMES SMITH, *Limekiln Lane, Greenwich.*"

In the early part of 1817 four engines were at work in the mines, two pumping water and two raising the ore; while a fifth engine was coining in the Mint at Lima. Trevithick believed that he could much improve the engines and the mining, and that it would be necessary and practicable to arrange for the construction of engines in Lima; for though death and dissension had caused difficulty, the authorities were still prepared to give him full power.

A strange defect in his character is evidenced in this letter. He wished his life to be insured for the benefit of his wife and family, but never thought of paying the yearly insurance premium, leaving it for his wife to pay, whom he had left, as far as he knew, penniless in England.

On his sailing from Penzance, he told his wife that he had paid the house-rent for a year in advance, mentioning the sum. At the end of that time a demand was made on Mrs. Trevithick for a year's rent, being a larger sum than her husband had mentioned as the proper rent. It turned out that Trevithick had taken and paid for the house at six-monthly periods, instead of yearly periods. It was in the same street, and but three or four houses from that occupied by the parents of the eminent Sir Humphry Davy.

A person pressed him for payment of a bill. Trevithick said, " Give me your bill," and writing on the bottom of it " Received, Richard Trevithick," handed it back to the claimant with " Now, will that do for you?" The payment of the life insurance obliged Mrs. Trevithick to part with her personal property, on which she had counted for support during her husband's absence. This inability to see the necessity of method-ical action, when working with others, and utter disregard for hoarded money, caused him to be a somewhat unmanageable partner, though his genius never allowed

MARKET, JEW STREET, PENZANCE. [W. J. Welch.]

him to sink; and in November, 1817, he wrote a letter, of which the following is an extract:—

"There are also nunneries beyond number, and in those places no male is ever suffered to put his foot. Through one of the most noted runs a watercourse, which works the Mint; and Mr. Abadia has repeatedly made all the interest he could to be admitted, for the purpose of inspecting it, but could never get a grant. The Mint belongs to our engine concern, and now coins about five millions per year. We have a contract from

Government for making all the coin, both gold and silver, which gives an immense profit; and as there must now be coined six times as much as before, I must build new water-wheels to work the rolls which we took with us from England. It was on this account that I wished to examine the watercourse for this purpose, without the knowledge of Mr. Abadia or anyone but Mr. Page and the interpreter, who always attends me. I walked up and knocked, in my blunt way, at the nunnery court door, *without knowing there were any objections to admit men*; it was opened by a female slave, to whom the interpreter told my name and business. Very shortly three old abbesses made their appearance, who said I could not be admitted. I told them I came from England, for the purpose of making an addition to the Mint, and could not do it without measuring the watercourse; upon which a council was held amongst them; very soon we were ordered to walk in, and all further nunnery nonsense was done away. We were taken round the building and were shown their chapel and other places without reserve.

" Uville knew nothing about the practical part of the engines, and Bull very little, therefore you may judge what a wretched state this great undertaking was in before my arrival; no one put any confidence in it, and believed it was all lost, together with five hundred thousand dollars that had been expended on it. The Lord Warden was sent from Pasco to offer me protection and to welcome me to the mines. They have a court over the mines and miners, the same as the Vice-Warden's Court in England, only much more respected and powerful. The Viceroy sent orders to the military at Pasco to attend to my call, and told me he would send whatever troops I wished with me. The Spanish Government and the Vice-king since my arrival are quite satisfied that the mines will now be fully carried into effect, and will do everything in their power to assist me. As soon as the news of our arrival had reached Pasco, the bells rang, and they were all alive down to the lowest labouring miner, and several of the most noted men of property have arrived here—150 miles. On this occasion the Lord Warden has pro-posed erecting my statue in silver. On my arrival Mr. Uville wrote me a letter from Pasco, expressing the great pleasure he had in hearing of my arrival, and at the same time he wrote to Mr. Abadia that he thought Heaven had sent me to them for the good of the mines. The water in the mines is from four to five strokes per minute.

" Tell the members of the Geological Society that Mr. Abadia is making out a very good collection of specimens for them, which will be sent by the first opportunity; and soon after I arrive at Pasco I will write them very fully."

After Trevithick's death, in 1833, casts were taken from the head, and busts presented to scientific societies were thankfully received, with the single exception of his near neighbours at Penzance, who, under the name of the Royal Geological Society of Cornwall, refused it.

Mr. W. J. Henwood, who had frequently drawn the attention of Cornishmen to Trevithick's engines, being about 1870 President of the Royal Institution of Cornwall, presented to it a bust of Trevithick, which was admitted within its walls.

In 1819 Mr. R. Edmonds forwarded to the 'Cornwall Gazette' news from Lima, from which the following is extracted :—

" We have much pleasure in stating that accounts have lately reached England from Lima, giving the satisfactory intelligence that our countryman and able engineer, Captain Trevithick, was in February last in good health, and superintending the rich and extensive mines of Pasco.

" Don Francisco Uville, a Spanish gentleman, having, with Don Pedro Abadia and others, formed a company to drain the mines of Pasco, unfortunately for Captain Trevithick, F. Uville was anxious to impress his countrymen with an opinion that it was *solely* owing to him that steam-engines were first introduced into the silver mines of South America; and notwithstanding the obligations he was under to Captain Trevithick, he sought every opportunity, soon after Captain Trevithick's arrival at Pasco, to oppose him, in claiming to have the direction of the mines.

"Captain Trevithick, knowing but little of the country, and disgusted with the treatment he received from Uville and the party he had formed against him, amongst whom was a gent who had lately arrived from England, retired from the concern, and proceeded on other important discoveries on his own account.

"Things remained in this state until August, 1818, when Uville met his death, in consequence of the cold penetrating air of the Cordilleras on coming out of the mines in a strong perspiration. Mr. Abadia and his friends were then under the necessity of soliciting the assistance of Captain Trevithick. On condition of his having the sole direction of the mines, he was prevailed upon to accept the situation which had been first most faithfully agreed he should have had; and when the accounts last left Lima in February, Captain Trevithick had been five months at the mines as the chief superintendent. The mines are represented as being in the most prosperous state, and likely to realize the sanguine expectations of the shareholders. Mr. Bull, an engineer from Chacewater, who left England with Uville, died at Pasco about ten months since."

When this was written, Trevithick had been two years in the country, and found the immense difficulties of the undertaking increased by jealousies and jobberies. Mr. Uville was no more, neither were Vivian or Bull; but one man remained alive out of the four who had sailed from England with the first cargo of machinery. In August, 1818, Mr. Abadia, who from the first was a leading authority, requested Trevithick to take upon himself the sole management of the mines, where he continued until April, 1819, as shown by the following extract from Captain Hodge's journal, supplied by my friend Mr. Charles Hodge :—

"The first time they met was at Lima, on the 26th April, 1819, at Dr. Thorne's; your father had just come down from the Cerro de Pasco mines. On the 8th May following, I find my father witnessed the hanging of three men for killing two of your father's men, named Judson and Watson."

Mr. W. B. Stevenson says :[1]—

"The Mint was established in Lima, in 1565. The machinery was formerly worked by mules, eighty being daily employed till the year 1817, when Don Pedro Abadia, being the contractor for the coinage, Mr. Trevithick directed the erection of a water-wheel, which caused a great saving of expense. In the year 1817 two Englishmen, sent from Pasco by Mr. Trevithick (who afterwards followed with the intention of working some of the silver mines in Conchucos), were murdered by the guides at a place called Puloseco. This horrid act was perpetrated by crushing their heads with two large stones, as they lay asleep on the ground. The murderers were men who had come with them from Pasco.

"I have heard Mr. Trevithick say, that on shaking hands with the men who work in those quicksilver evaporating rooms, drops of quicksilver show themselves at the fingers' ends, and that the workmen wearing shoes take them off before leaving the work, to pour out any quicksilver that had oozed through the pores of the skin, which had been respired in the floating state of vapour. The men so employed fell a sacrifice in twelve or eighteen months."

Trevithick's experience in applying the force of running streams was turned to good account in giving an economical helpmate to the steam-engine then at work in the Mint. Miers says :[2]—

"Another instance occurs in the unfortunately ruinous result and lamentable ill-treatment of the persons engaged in the attempt to introduce European improvements and British machinery into the great silver-mining district of Pasco (Chili), in which was engaged one of our most celebrated engineers, a most able mechanic, to whom the grand improvements in our

[1] See 'Historical and Descriptive Narrative of Twenty Years' Residence in South America,' by W. B. Stevenson, published 1842.
[2] See Miers' 'Travels in Chili and La Plata,' published 1826.

Cornish mines are chiefly indebted—I mean Mr. Richard Trevithick. Trevithick was induced to furnish the machinery at
an expense of 3000l. sterling, upon condition of being admitted
a partner in the amount of 12,000 dollars in the joint stock of
the company, and entitled to a share corresponding to the
capital employed. This share was calculated at a fifth. Trevithick, before he embarked for Peru, divided his interest in
the concern into 320 shares, each representing 38 dollars, and
these were sold in the market for 125l. sterling each ; some few
were sold for 100l. cash. The success of the engines gave to
some of the persons interested much confidence, who conceived
they could now do without the management of the ingenious
Trevithick. Every possible obstacle was therefore thrown in his
way by those who, from motives of jealousy, wished to get rid of
him. The persons to whom Trevithick's and other shares had
been sold in London, sent out to Lima an agent, whose duty it was
to look after their interests in the concern ; but as it was found
a much larger sum would be necessary for carrying the enterprise into effect than had been calculated, a collision of interests
took place ; complaints were made on all sides as to the delays
and expenses which those who did not comprehend the almost
insurmountable difficulties of the undertaking attributed to
mismanagement and carelessness. The greatest share of oppro-
brium fell unjustly upon Trevithick, who, being a man of great
inventive genius and restless activity, was at length completely
disgusted, and retired from the undertaking. He left Pasco,
although Abadia offered him 8000 dollars per annum, together
with all his expenses, if he would continue to superintend the
works ; on no conditions would he consent to contend with the
jealousies and ill-treatment of the persons with whom he had
to deal. He soon after entered into speculations with some of
the miners at Conchucos, for whom he constructed grinding
mills and furnaces, with the view to substitute the process of
smelting for that of amalgamation in silver ores, in which vain
pursuit he became a considerable loser."

" MY DEAR SIR,   " BODMIN, *November 3rd*, 1869.
        "Forty-seven years are now passed since I had the
great pleasure of meeting your father in Peru, and I have a

vivid remembrance of the gratification afforded to my messmates when he came to dine with us on board H.M.S. 'Aurora,'
then lying in Callao. I was then a lieutenant of that beautiful
frigate, and was introduced to your father by Mr. Hodge, of St.
Erth, with whom I had become acquainted in Chili. I remember your father delighting us all on board the 'Aurora' by his
striking description of the steam-engine, and his calculation of
the 'horse-power' of the mighty wings of the condor in his
perpendicular ascent to the summit of the Andes. Your father's
strong Cornish dialect seemed to give an additional charm to
his very interesting conversation, and my messmates were most
anxious to see him on board again, but he left shortly after for
the Sierra.

"The Pasco-Peruvian mines were those which your father
was engaged to superintend before he left England, and he had
actually managed, by incredible labour, to transport one or two
steam-engines from the coast to the mines, when the war of
independence broke out, and the patriots threw most of the
machinery down the shafts. This fearful war was a deathblow
to your father's sanguine hopes of making a rapid fortune.
About a year after this terrible disappointment (I think in
1822), the 'San Martin,' an old Russian fir frigate, purchased
by the Chilian Government, sank at her anchors in Chorillos
Bay, ten miles south of Callao, and your father entered into an
engagement with the Government in Lima to recover a large
number of brass cannon, provided that all the prize tin and
copper on board which might be got up should belong to him.
This was a very successful speculation, and in a few weeks your
father realized about 2500*l.* I remember visiting the spot with
your father whilst the operations were carried on, and being
astonished at the rude diving bell by which so much property
was recovered from the wreck, and at the indomitable energy
displayed by him. It was Mr. Hodge, and not I, who then
urged in the strongest manner that at least 2000*l.* should be
immediately remitted to your mother. Instead of this, he
embarked the money in some Utopian scheme for pearl fishing
at Panama, and lost all!

"I had the honour of dining with Lord Dundonald on board

the crazy frigate ' Esmeralda,' which carried his flag in Callao
Bay, but I never heard of the gallant conduct of your father
in swimming off to his ship and advising him of an intended
assassination. I fancy that this must have occurred before I
came on the station, probably in 1820, or 1821.

" Believe me,

" My dear Sir,

" Very sincerely yours,

" JAMES LIDDELL."

Trevithick's floating caissons for the sunken ship of
Margate Bay in 1810[1] were similarly applied in 1821
in the Bay of Callao. In Lima he became acquainted
with Lord Dundonald, whom he warned of a plot on
his life, discovered in his friendly intimacy at the
residence of President Bolivar. Those two remarkable
Englishmen were alike in their daring inventiveness,
and not unlike in face and person.

We have traced Trevithick's steps from his landing at
Lima in 1817 to the destruction of the mine machinery
by the civil wars, and his departure from there about
1822. But one link in the chain has been nearly
lost. During some portion of those five years he
visited Chili, and set to work mines which are still
producing large and profitable quantities of copper.
The late Mr. Waters, an eminent Cornish miner, who
for many years managed some of these mines in the
neighbourhood of Valparaiso, said that Trevithick's
name was better known to the miners there than to the
miners in Cornwall. This statement was made in the
Dolcoath account-house at a public meeting, the speaker

---

[1] In 1834 the writer was employed
at a marine engine works in London,
and made working drawings for a
scheme of Lord Dundonald's, who expressed great pleasure in meeting the
son of his old friend.

and the writer being both on the committee of management.

Simon Whitbarn, of St. Day, informed the writer that at Copiapo and at Coquimbo he had seen large heaps of copper ore, apparently unclaimed, which the people said had been raised by Don Ricardo Trevithick. About 1830 a miner, returned from South America, made a claim for wages for watching mineral left behind by Mr. Trevithick.

To further illustrate this history, we have a report written by himself:—

*" Memoranda regarding the Copper and Silver Mine of * * * ***

" In 1814 an arrangement was made between the miners of Peru and myself for furnishing them with nine steam-engines and a mint, to be executed in England and erected in the mines of Pasco; and in October, 1816, I sailed from England for that country, for the express purpose of taking the management of those mines and erecting the machinery, being myself a large proprietor of the same. The Government of Peru was at that time subject to old Spain, under the immediate superintendence of a Viceroy. The machinery having been erected, and its sufficiency for the intended purpose of draining the mines having been proved to the satisfaction of all parties, there was granted to me a special passport by the Viceroy, for the purpose of travelling through the country to inspect the general mining system, and to make the native miners acquainted with the English modes of working. In return for which Government conceded to me the privilege of taking possession for my own benefit and account of such mining spots as were not previously engaged. In this way I travelled through many of the mining districts, and although I met with several unoccupied spots which would have paid well for working, yet, being a considerable distance inland, and requiring more capital to do them justice than I could then advance, I abandoned for the time all ideas of undertaking them.

" To this, indeed, there was but one exception, and that was a
copper and silver mine, the ores of which are uniformly united,
in the province of Caxatambo.

" When the patriots arrived in Peru, the mine was deserted
by all the labourers, in order to avoid being forced into the army.
In this state it remained for a considerable time; but on the
Spaniards retreating into the interior, I recommenced working;
and to secure my right to this mine under the new Government
I at the same time transmitted a memorial and petition to the
established authorities, accompanied by a plan and description
of the mine, the result of which was the formal grant, as exhibited in the Spanish document now in your possession. It
was not my good fortune to be allowed to follow up my plans,
which almost warranted a certainty of success. I had scarcely
commenced a second time when the Spaniards returned, and
everyone again was obliged to fly. The country, as is well
known, continued for a long time in a most distracted state, and
I was ultimately compelled to quit that part of Peru, robbed of
all my money, leaving everything behind me, miners' tools and
about 5000l. worth of ores on the spot ready to be carried to
the shipping port. Numerous as my misfortunes had been in
Peru, and heavy as my disappointments, I felt none so sensibly
as this, because it was an enterprise entirely of my own creation,
and so open to view that I was enabled to calculate at a
certainty the immense value contained within the external
circle where the copper vein made its appearance in the cap of
the mountain, and to be obtained without risk or capital.
However, revolution followed revolution, and the war appeared
to me to be interminable. Even Bolivar's arrival at Lima made
it still worse, for he forced me into the army, with my property,
which is not paid to this day, to the amount of $20,000; and
at his urgent solicitations, disgusted as I was with what I had
seen and suffered in Peru, I determined on quitting it for a
time at least, and on visiting Colombia. Being at Guayaquil I
first heard the name of Costa Rica and its recently-discovered
mines, and having no doubt of the authenticity of my information, I immediately proceeded thither instead of going to
Bogota to carry Bolivar's orders into execution, not having

been paid. This short digression you will excuse, as it points to the causes of my separation from a property of so much value, as I consider the mine of * * * * * Thirty years ago the neighbourhood of * * * * was famous for its silver mines. At the foot of the copper hill, on a fine stream, are two sets of works on a most extensive scale, which were carried on on account of the Spanish Government. The silver was found in lead veins, which are very large and numerous all around. The soil is very rich, and the climate as good as any in the world, wheat and Indian corn both growing round the mountain. Provisions and wages are low, the latter 1s. per day, and there are about 20,000 inhabitants within three miles. Wood for smelting and other purposes is abundant on the spot.

"* * * * is * * leagues from Lima; the port of * * * * where the ores are to be shipped, is 37 leagues north from Lima; and * * * * copper mine * * leagues back in the country east from this port, a good road for mules and plenty of them. The miners contracted with me to break the ores and deliver them at the surface for 4l. per ton, which was double what I ought to have paid them; the farmers likewise contracted to carry the ores to the port at the same rate, which comes to sixpence a league for each mule cargo. But even at present wheel-carriages might travel over a large proportion of the road, and a small outlay would make it a carriage-road the whole distance, and then the expense of carriage would be diminished more than one-half. Taking it, however, at what it cost me, the whole expense on the ores delivered on board would not amount to 9l. a ton, and as I conceive the freight to England would not exceed 4l. a ton, the total cost would be 13l., but say 15l. a ton. Its value in England would be above 80l. a ton. At the time I worked I intended to have sent 300 tons of ore to England, for in the then disturbed state of the country it would not have been prudent to risk myself on smelting works. I think it will ultimately be found preferable to smelt on the spot, but the course I should recommend in the meantime would be to send out two practical miners to direct and superintend the natives, who ought to be employed by contract to break and raise the ores and deliver them on board.

In that case no erections whatever would be wanted; nothing but about 70*l*. worth of labourers' tools.

> " I remain, Sir,
>
> " Your very humble servant,
>
> (Signed)   " RICHARD TREVITHICK."

The foregoing undated report was written after his return to England from South America. The Viceroy granted him a special passport through the country, that he might give general instructions to the workers of mines, with the right to claim any mineral spot for his own working not under grant to others.

He often spoke of his discovery and working of the great vein of copper ore in Caxatambo, estimated to contain copper worth twelve millions sterling, the working of which was prevented by the frequent revolutions and unsettled government of the country; and of residing for months with Bolivar, at that time the Republican Governor of Peru.

Bolivar's cavalry were short of fire-arms. Trevithick invented and made a carbine with a short barrel of large bore, having a hollow frame-work stock. The whole was cast of brass, stock and barrel in one piece, with the necessary recess for the lock; the bullet was a flat piece of lead, cut into four quarters, held in their places in a cartridge until fired, when they spread, inflicting jagged wounds. He was obliged to serve in the army, and to prove the efficiency of his own gun. He was never a good shot, nor particularly fond of shooting; and, after a long time, Bolivar allowed him to return to his engineering and mining. Scarcely had he got to work again when the Royal Spanish troops, getting the best of it, overran the mines, and drove Trevithick away penniless, leaving 5000*l*. worth of ore behind him ready for sale.

The 300 tons of ore, valued at 24,000*l.*, never reached England; and the writer, who was to have returned to Peru in the ship that had been engaged to convey it, lost the chance of being a youthful traveller in foreign lands.

Trevithick left Lima about 1821 or 1822, for Bogota, in Colombia, on a special mission for Bolivar. On his way, putting in at Guayaquil, he heard of rich mines in Costa Rica, and thinking they would pay better than Bolivar's promises, he threw up his engagement and made for the new venture. It was probably at Guayaquil that he met Mr. Gerard, a Scotchman of good family and education, then sailing on the Pacific coast as a speculator.

Since Trevithick left the mines of Cerro de Pasco, more than one English adventurer has attempted to work them. At the present time they are in the hands of a large company, and are thus spoken of in the 'Cornish Telegraph' of May 10, 1871 :—

" Cerro de Pasco and its Silver Mines.

"This place, in the Republic of Peru, is situated on the top of the Andes, on the eastern side of the Western Cordillera. It stands about 15,000 feet above the sea level, and is said to be one of the highest, if not the highest, inhabited place of importance in the whole world.

" From Callao to here is a distance of 160 miles, but, in consequence of the rapid ascent in such a comparatively short distance, it is considered a quick journey if mules make it in six days; it more frequently takes them a week, and at times, during the season of snow and rain, the pampas, which are the table-lands of these mountains, are impassable for several days together.

" The town of Cerro de Pasco, which at present numbers 10,000 souls, is of no small importance, considering its great altitude and inconvenient distance from the coast, but it lacks

order and design in every part. The streets are crooked and
uneven; and the·houses are stuck about anywhere and everywhere, with the greatest display of uneducated taste that I
have ever before witnessed; moreover, it would be difficult to
find another such place so equally dirty.

" It rains and snows on these heights with not much cessation
for about six months in the year, and in what is termed the dry
season there are also frequent falls of snow. Furthermore, water
boils at 180° Fahr. instead of at 212°, as with you; consequently
it requires six minutes to cook an egg.

" The majority of the inhabitants are a low type of Indians,
who are small in stature and mind, but are large in cunning,
and have exceedingly plain features—not possessing the
slightest trace of the noble features and bold simplicity of the
Indians of the North.

" Any person acquainted with minerals and mining coming
up to Cerro de Pasco would fancy that the whole town was
built on the back of one huge lode; go wherever one may,
through the streets, or on the outskirts of the town, and even
up to the slopes of the hill surrounding it, he finds it to be all
lodestuff everywhere; its composition is what we Cornish
miners generally term an iron gossan.

" The greater portion of this mineral spot is parcelled out
into setts or grants, which consist of pieces of ground 60 yards
in length by 30 in width, giving to the place no less than
664 mines. At present there are no more than seventy-eight of
them at work, and only sixty-three of which are producing ore,
and the united returns amount to 2,000,000 oz. of silver per
annum. Owners or compan es have roads leading down to
their mines, formed of steps cut out in the rock, dipping at
angles varying from 30° to 50°. When you have descended to
the depth of the mines, the levels or holes leading to many of
them are so small that one has to drag himself along snake
fashion until he reaches the main excavation. The miners
break down the silver ore with pointed bars of iron, and then
shovel it into bags made of hide with the shoulder-bone of some
animal , after which the stuff is carried to surface on men's and
boys' backs.

" When all the mineral has been extracted there remains an immense excavation, and in consequence of the roof not being properly supported with timber, one risks his life in entering it. Heavy falls of rock frequently occur, and by which means a vast number of persons are annually killed. One day in the last century, at the mines of ' Matagente' (which word means killed people), which are situated in the rising ground on the northern side of the town, while a great number of men and boys were at work, the roof of one of these immense chambers, consisting of many thousands of tons, fell in without giving the least warning, and 'in the twinkling of an eye' the souls of 300 Peruvian miners rushed into the presence of their Redeemer. Their bodies have never been exhumed, and their shattered bones, still remaining, will bear evidence of the catastrophe to future explorers. An adit has been driven through the district, beginning at the Lake of Quiulacocha on the south-west, and terminating at the mines of Ganacaucha on the north. The entire length of the adit, including its branches, is about 3 miles, and its average depth from surface 50 fathoms. Three perpendicular shafts, situated at about 600 yards apart, have also been sunk from surface to a short distance below the adit.

" The whole of the machinery for the mines in question, which is being made and dispatched by Messrs. Harvey and Co., of Hayle, Cornwall, consists of four steam pumping engines, six boilers, four iron main beams, four balance ditto, and also a sufficient quantity of 24-inch pit-work for both shafts. No single piece of all this cumbrous machinery must weigh more than 300 lbs., in consequence of its having to be transported on the backs of mules from the coast to this mountainous region. Although the main distance is no more than 160 miles, these beasts with their burdens have to climb an altitude of 15,000 feet before they reach their destination. Moreover, the passes in ascending the Andes and Cordillera can only be correctly imagined by experienced travellers. Some of the defiles are not much wider than a sheep-path, and with a thousand feet below you a roaring cataract, and thousands of feet above you snow-capped overhanging mountains, looking so dreadful that

the awe-struck stranger in the pass fears that the next peal of
thunder will cause them to topple."

MULE TRACK FROM LIMA TO CERRO DE PASCO.   [W. J. Welch.]

"I observe in a paper which is now before me, entitled 'The
Introduction of the Steam-Engine in the Peruvian Mines, by
Richard Trevithick, in 1816,' that when Captain Trevithick

arrived at Lima on board the ship ' Asp,' with sundry small engines for the draining of the mines of Cerro de Pasco, he was immediately presented to the Marquis de Concordia, then Viceroy of Peru, was most graciously received by the most flattering attention of the inhabitants, and subsequently the Viceroy ordered the Lord Warden of the mines to escort the great man with a guard of honour to the mining district. In contrasting the two epochs, that of Trevithick in 1816, with this of Wyman and Harrison in 1871, one is led to exclaim that there were *gentlemen* in Peru in 1816, and they gave unto Cæsar that which belonged to Cæsar." [1]

The same newspaper, on the 9th November, 1870, stated :—

" The ' Bride ' sailed from Hayle on Thursday with a portion of the machinery made by Messrs. Harvey and Co., of Hayle, destined for Cerro de Pasco, in Peru. The work comprises four 37-inch cylinder pumping engines ; no part to weigh more than 300 lbs."

To enable the parts to be reduced in weight, each steam-cylinder was made of thirty-seven different pieces. The mechanics of Trevithick's time could not make a steam-cylinder in parts ; therefore his difficulties in designing and conveying the machinery were ten times greater than they would be in the present day, and necessitated the extreme simplicity of his engines. His residence with the Peruvians from 1816 to 1822 taught them the use of high-pressure steam-engines in their mines ; and indirectly heralded the advent of the steam-horse, now as familiar to them as to the residents in many English towns.

---

[1] In ' Mining Journal,' W. R. Rutter.

CHAPTER XXIII.

COSTA RICA.

" My dear Sir,

"In the month of June, 1822, I disembarked in the port of Punta de Arenas, in the Gulf of Nicoya, the only one corresponding to that province at present in use on the Pacific side. My object was to dispose of a cargo of cotton which I had brought from Realejo, and to purchase sugar in return. Circumstances, not necessary to mention, and the loss of the small vessel with which I was trading on the coast, caused me to remain in Costa Rica. Its name implies a very early conviction of its natural opulence; it is certain that gold and silver abounded among the Indians at the period of its conquest by the Spaniards. It was at one time a favoured and flourishing agricultural colony, but from various causes sank into neglect. Such was the apathy, both of the Government and of individuals, that the very existence of the precious metals in the country had been almost entirely forgotten. In the end of 1821, a poor man, Nicolas Castro by name, opened the first gold mine known in Costa Rica since the conquest, and his success soon induced others to try their fortunes; with fortunate results, in a few months a mining district sprang into being.

"A gentleman of the name of Alverado constructed at a very considerable expense what is called an Ingenio, consisting of various edifices for depositing the ore, machinery driven by water for grinding it and afterwards blending it with quicksilver for amalgamation.

"When I landed in June, 1822, only five or six mines had been discovered, but in January 1823, when I left the country, I cannot pretend to enumerate those in a state of progress and of promise. It is not only in the mining part of the business

PLATE 14.

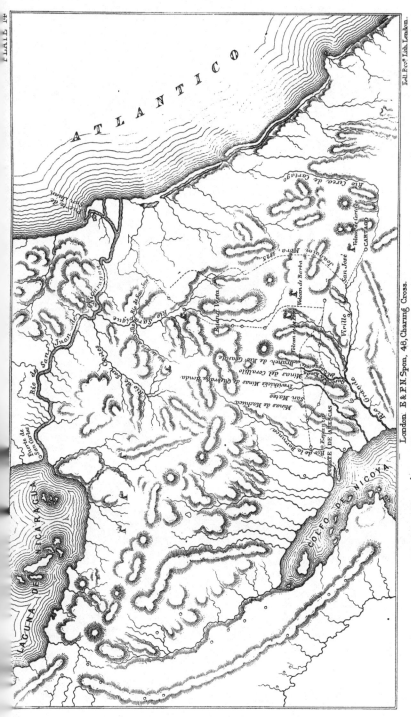

Edd. Pro^s Lith. London.

London: E & F N Spon, 48, Charing Cross.

TREVITHICK'S ROUTE ACROSS THE ISTHMUS OF COSTA-RICA.

that the want of skill is prejudicial to the result. It is imperfectly ground, for instance, and consequently cannot be brought into that intimate contact with the quicksilver which is necessary to perfect amalgamation. The machine for grinding is very simple: a large flat stone, like a mill-stone, is made to revolve upon its fellow by an ox or mule power. The poorest people reduce it to powder by manual labour, in the same way as they grind corn preparatory to baking it into cakes. Alverado's machine promised to be a great acquisition. The grinding was facilitated by a little water; when the ore is judged to be sufficiently well ground, a portion of quicksilver is thrown in by guess, and the motion of the machine continued until the union of the metals is supposed to be complete; the whole is then removed into large wide-mouthed conical-shaped wooden vessels. In these receptacles it undergoes repeated washings, by stirring occasionally round, and afterwards communicating to the vessel a swinging or half-rotary motion, by which a quantity of the water, having the earthy particles suspended, is driven over the edges; the amalgamated mass naturally sinks to the bottom, and at last remains tolerably clean.

"The next step is the recovery of the quicksilver by distillation, after which the gold is melted in a crucible and run into ingots. The coasts are hot, and from the luxuriant vegetation that everywhere abounds, emit, as in all situations of the kind, febrile miasma in abundance when acted on by heat and moisture; but black vomit is unknown, and all the fever cases I have seen have been of the remitting and intermitting, free from character of malignancy. As the ground begins immediately to spring from the coast, and does so indeed very rapidly, a few miles takes us beyond the region of even these slight fevers, and as we continue ascending to the central table-land, a climate is encountered that may vie with any in the world for benignancy and beauty. We there meet with the fruits of the torrid zone, and near them the apple and the peach of Europe. The orange tree is in bearing the whole year. As in all situations within the tropics, it has a proper rainy season, but it is less inconvenient and disagreeable than might be expected, for it seldom rains two days in succession, and when it does, is invariably

succeeded by an interval of fine weather; for the most part
every day presents a few dry hours.   The mines are situated on
the ridges of the Cordillera, which without presenting snow-
covered peaks, attain, nevertheless, considerable elevation.
The clouds, constantly attracted by those high summits, render
the rainy season more severe in the mining district than in the
plains.   The greatest inconvenience was from the snakes, which
in those solitary jungles, now first invaded by man, are very
numerous and many of them venomous.   Provisions are cheap
and excellent.   In short, there is but one fault I find with the
country, and it is a great one, I mean the frequency of earth-
quakes.   "J. M. GERARD."

MEM. IN MR. GERARD'S WRITING.

"Illustrations of the Map.

"Though the plans and sections explain themselves, a few
observations will not be misplaced.   The deep adit for the Coralillo would be 600 yards, that for Quebrada-honda 400 yards,
and besides serving as drains would form admirable roads for
conveying the ores into the vale where the stamps must be
erected.

"The veins would be worked upward from the adits, and
thus no expense would be incurred for ages to come in lifting
either water, ore, or rubbish to the surface.   Padre Arias Mine
is an exception, requiring a powerful water-wheel, or an
hydraulic pressure-engine, for which there is a fine fall of
water of 135 feet.   The mines in Quebrada-honda are those
in which an interest has been procured.   Captain Trevithick
has an interest in the mine of Coralillo; the great watercourse
is also his.

"It will be seen by the plan that there are 75 fathoms fall
to the point where his present mill is situated, and other
75 fathoms to the junction of the rivers of Quebrada-honda and
Machuca.   The whole length does not amount to two miles,
within which it is estimated that sufficient power may be com-
manded to stamp 500,000 of quintals annually.   To bring it up
to that pitch, the waters of Machuca must be brought to join

those of Quebrada-honda at Trevithick's mill, and then 40 tons of water per minute could be delivered in the dry season."

Extracts from a report by Trevithick and Gerard in 1827 :—

" This map consists of several distinct parts. The middle part shows the mining district, the present dimensions of which are small, the length being hardly four miles, breadth from two to three, and the superficial extent from eight to ten miles. The upper part of the plan is a section of the north ridge, called Quebrada-honda, and shows the line of the proposed adits. The lower part in like manner exhibits the south ridge, called Coralillo. The map further shows the inclination or gradual fall of the ground along the valley, and of the streams by which the mills are driven.

"The canal is likewise shown 5000 yards in length, by which the rivers of Machuca would be brought to join that of Quebrada-honda.

" Castro's mine is situated on the southern ridge, and was the first mine worked to any extent. There the veins are very large ; in fact, from the manner in which a number of horizontal veins are seen falling into the perpendicular or master vein, the great body of the mountain would appear to consist of lodes. This mass of ore is in general rich. It has been worked open to the surface, somewhat like a quarry, so that it is not difficult to calculate in cubic feet the quantity that has been excavated. The mine is supposed to have yielded in the course of the last six years gold to the value of 40,000*l.*, and by measuring the excavations it would appear that this amounts to, on an average, one ounce of fine gold to every ten or twelve quintals of ore. In 1821 the existence of silver was only imagined. In 1823 it was fully ascertained. Ever since 1824 it has constituted a small but constant portion of the produce of Quebrada-honda, and in 1827 it was decidedly evinced in Coralillo. The discovery of gold in Coralillo led them to work in Quebrada-honda, where they found both gold and silver, and the discovery of silver in Quebrada-honda, by strengthening the expectation of it in Coralillo, led in its turn to the discovery of silver there.

In Quebrada-honda they only work on the ground in the immediate vicinity of the stream, and that in the most imperfect
manner; but great light has been thrown on the value of the
ores on this spot and in the district generally by the progress
made in working what is called Padre Arias Mine, which takes
its name from an ecclesiastic who first worked it. This mine is
situated in low ground near the verge of the stream, and was at
first only worked for gold. There were soon, however, indica-
tions of silver, which increased progressively in sinking, till at
the depth of only 10 yards the influx of water exceeded the
means of draining, and the works under water-level were neces-
sarily abandoned, at a time when ores were yielding upwards of
200 oz. of silver to the ton, a striking proof of the tendency of
silver ore to improve in this district as the depth increases.

"Mr. Richard Trevithick, that eminent Cornish miner and
engineer, so well known for his inventions, and particularly for
the high-pressure steam-engine and the drainage of the Pasco
Mines in Peru, when unfortunately civil war burst out in Peru,
and the Royalists, considering those engines as the main instrument for supplying money to the Independents, rendered
them useless by destroying or carrying off some of the most
important pieces.

"Mr. Trevithick having heard favourable reports of the
mining district we are now describing, soon after repaired
thither, and was so fully impressed with its value and importance that he made an extensive contract for different properties,
and resided in the country for four years.

"He is now in England ready to give explicit answers to any
inquiries that may be made as to the mineral wealth of Costa
Rica, and the extraordinary facilities afforded by its position
and natural advantages. An estimate has been made for establishing a complete mining concern in Costa Rica, with houses,
iron railroads, stamping mills, &c., so as to raise, stamp, and
bring into refined gold the produce contained in 250,000 tons
of ore per year.

"The result of six years' experience shows that the following
list of machinery and tools with a few miles of railroad would
be sufficient. The communication with the mines being satis-

factorily established by the route of the port of San Juan de Nicaragua and the river Serapique, the materials would be sent by the Atlantic at very much less cost than by around Cape Horn.

"It is situated within 14 leagues of the Pacific Ocean and 30 leagues of the Atlantic, in a mountainous district intersected by deep valleys or ravines. The mountains are covered with wood fit for fuel, mining, architecture, and machinery. There is a population of 50,000 inhabitants within one day's journey of the mines. The climate is perfectly salubrious, provisions of all kinds remarkably cheap, labourers' wages from four to five dollars per month. The mines secured are freehold property, and with one exception are unencumbered by tribute or native partners. The attention of Government and of individuals has recently been directed to the discovery of a road from the interior to the river Serapique, which, rising in the high lands of Costa Rica, pursues a northerly course and joins the San Juan about 10 leagues above the harbour of that name, being itself navigable for about 12 leagues above the junction. The opening of this road is a matter of much importance to Costa Rica in a general point of view; the port of Matina being always bad and impracticable during the prevalence of northerly winds; that of San Juan being, on the contrary, capacious, easy of access, and at all times perfectly secure. The distance is much the same as by the way of Matina. Several expeditions have been undertaken with a view of exploring an eligible road to the highest navigable point of Serapique, and although as yet none fit for mules has been discovered, the results of the experiments justify the expectation of success. Individual enterprise is active in the attempt, and Government has wisely offered a reward to successful speculators.

"Captain Trevithick and Mr. Gerard, with a particular view to the enterprise now under consideration, and after considerable risk and labour, succeeded in laying down the navigable head of the Serapique and in throwing such light on the intervening tract as will be of great assistance to future adventurers. They ultimately constructed a canoe in which they sailed down to the port of San Juan."

Plate XIV. shows Trevithick's route across Costa Rica.

A memorandum in Trevithick's writing, apparently a diary, says:—

" From where we returned our mules to the place where we commenced to make our rafts and boat was eleven days' journey, a distance of 50 or 60 miles. The first and second days after parting with the mules we passed some soft ground, with three or four rivulets of water in narrow vales, about 10 miles on the side of the decline of the high ridge on our left. It could easily be made passable for mules, as the bad places where they could not travel did not exceed two or three miles; and had we kept a little more to the left above the soft ground, probably they could have passed. The next bad place was about a mile after the second pass across the San José River, being a very deep and abrupt vale. Had we never passed the San José River, but left it on our right hand, the road would have been much shorter, and we should have avoided this deep vale, and also the three other vales, and their three rivers of Montelegre, Juan Mora, and Ajerbi. They were, however, small, not more than half the leg in water, which is a proof that their source was not above 10 miles off and must have originated in the side of the high ridge on our left. None of the vales were impassable to mules, except that between the second passing of the river San Jose and the river Montelegre, which was about a mile, and might be made passable for mules by a diagonal road to be made in the side of the hill a little higher up.

" Only five or six miles of road would require to be made for mules on the whole of the way we came, to where the river Serapique is navigable. We observed that we should have avoided those vales by passing a few miles more to the left, where we saw one continued high ridge running from the highest ridge of the continent, commencing at the volcano and terminating in a point near to where the Serapique River is navigable.

" On a regular decline for perhaps 7000 or 8000 feet in height, down to near sea-level, which would in that distance have given a fall of about half an inch in a yard, four men in ten days would make, I have no doubt, this ridge passable for mules on a regular descent to where the Serapique River is navigable. I have no doubt if we could have spent one week more on our journey we might have passed mules the whole distance with us. To carry machinery from where the Serapique is navigable to the mines is about one-third farther than from the port of Arenas on the south, on which the carriage is two dollars per mule load; three dollars might therefore be charged per mule from the Atlantic side, a much less cost than by way of Matina, or by going around Cape Horn. It would give a speedy communication and a great accommodation to the province of Costa Rica, which I doubt not would gladly contribute to its making.

"The mining district occupies the mountain of Aquacate, nearly equidistant from the port of Punta de Arenas, in the Gulf of Nicoya, and from San José, the capital of the state, about 14 leagues from the former and 12 from the latter. The high road passes through the centre of the district.

"The chief outlay after paying for the mines would be for erecting stamping mills and making railroads."

This broken information barely gives an idea of the importance of the Costa Rica mines, or of what Trevithick did between the time of his landing on the Pacific shore, about 1822, and his leaving the mines on his search for a new route over the Cordillera to the Atlantic shore, about 1826 or 1827. Judging from the rough map on which Trevithick has marked his line of travel across the isthmus, the mines of Machucha, Quebrada-honda, and Coralillo, were inland from the Gulf of Nicoya, on the Pacific, some forty or fifty miles, the latter mine having its water shed into the Rio Grande, while the two other mines, not far off, opened into the Quebrada-honda River. The central high ridge of the Cordillera was between the mines and

the Atlantic; indeed the mines are on high ground at
the foot of volcanic mountains. San Mateo seems to
have been the place of importance near the mines, and
probably a well-known mule-track was in use through
the mountain ridge to San José, the capital, once
numbering thirty thousand inhabitants; but this line
failed to reach a good port on the Atlantic coast. The
travellers, therefore, abandoned the known track, and
turning to the left, made their way between the volcanic
peaks of Potos and Barba, hoping that on the eastern
slope of the Cordillera navigable rivers would be found
either to the Atlantic or to the San Juan de Nicaragua,
which joined the Atlantic at the port of San Juan. It
was probably at this volcanic ridge that the precipitous
road obliged the mules to be sent back. The track was
then due north, towards Buona Vista, below which the
river Serapique took its rise, running into the river San
Juan. Where they crossed this river was fifty or sixty
miles from where the mules had left them. Trevithick
marked the river-crossing with a steamboat, indicating
its navigability; but the writer infers that it had so
much of the mountain torrent about it, that the
travellers took a line still through unexplored country
towards the port of San Juan, on the Atlantic, for
the track and the description show that the river San
José was crossed, and also another river running to
the Atlantic. They probably were stopped by swamps
on approaching the San Juan, and retracing their steps
to the Serapique, constructed rafts or canoes, and after
hairbreadth escapes sailed down it to the junction with
the San Juan, and down the latter to its junction with
the Atlantic at Port San Juan, or Greytown.

Eleven days were passed from the parting with the
mules near the crossing of the highest ground, from

whence they saw a continuous ridge, commencing at the volcano and terminating near to where the Serapique is navigable on a regular decline for perhaps seven or eight thousand feet down to near sea-level, giving a fall for the whole distance of about half an inch in a yard, or in railway parlance 1 in 70; for this was what was in Trevithick's head, that his steam-horse should carry where the mule could not, and that miners and machinery should be so taken to his mines from the Atlantic, giving those who chose an opportunity of continuing their railway journey to the Pacific.

The writer has heard Trevithick describe the excursion as lasting three weeks, through woods, swamps, and over rapids; their food, monkeys and wild fruit; their clothes, at the end of the journey, shreds and scraps, the larger portion having been torn off in the underwood.

Mr. Thomas Edmonds also listened to Trevithick's narrations, some of which he gives in the following:—

"In 1830 I frequently saw Trevithick at the house of Mr. Gittins, at Highgate, a schoolmaster, with whom were two boys that had accompanied him from Costa Rica, called Montelegre. Before Captain Trevithick no European had adventured on or explored the passage along the river from the Lake Nicaragua to the sea. In the adventure he was accompanied by Mr. J. M. Gerard, a native of Scotland; two boys of Spanish origin going to England for their education; a half-caste, as servant to Mr. Gerard; and by six working men of the country, of whom three went back, after helping to remove obstructions in the forest through which the first part of the journey was undertaken. The risks to which the party were exposed on their passage were very great: they all had a narrow escape from starvation, one of the labourers was drowned, and Captain Trevithick was saved from drowning by Mr. Gerard. The intended passage was along the banks of the river. To avoid the labour of cutting through the forest, the party determined to construct a raft, on which they placed themselves, their provisions, and utensils; after a passage of no long duration they came to a rapid, which almost overturned their raft, and swept away the principal part of their provisions and utensils. The raft, being unmanageable, was then stopped by a tree lying in the river, with its roots attached to the bank; on this tree three of the passengers, including Captain Trevithick, landed, and reached the bank; this was no sooner done than the current drove the raft away from the tree, and carried it, with the remaining passengers, to the opposite bank, where they landed in safety, and abandoned the raft as too dangerous for further use. The next object was to unite the party again into one body. The three left on the other side of the river were called upon to swim over: one of the men swam over in safety, the next made the attempt and was drowned, the third and last remaining was Captain Trevithick, who was either unable to swim or could swim very little. In order to improve his chances of safety, he gathered several sticks, which he tied in a bundle and placed under his arms; with these he plunged into the stream; but the contrivance of the bundle of sticks afforded him very doubtful assistance, for the current appeared to seize the sticks and whirl him round and round. He, however, finally reached within two or three yards of the bank in a state of extreme exhaustion. Mr. Gerard going into the water himself and holding the branch of a tree, then threw to his assistance the stem of a water-plant, holding one of the extremities in his own hand. It was not until the fourth time of throwing that Captain Trevithick was able to seize the very extremity of the plant (which was leaf) in his fingers; on the strength of the leaf his life on the occasion was dependent. It was determined to give up any further idea of using a raft on the river, and to continue their journey along the banks of the river. For subsistence for the remainder of their journey they had to depend on the produce of one fowling-piece and a small quantity of gunpowder; after a few days the gunpowder got wet by accident, and in the attempt to dry it, it was lost by explosion. The party finally arrived in a state of great exhaustion at the village, now the considerable port of San Juan de

Nicaragua, or Greytown; and shortly after their arrival a small vessel arrived, which conveyed the party to one of the West India islands.

" Upon one occasion Captain Trevithick was called upon to act in a novel capacity, that of a surgeon. An accident happened to a native engaged in working an engine erected at a place distant about two hundred miles from Lima, by which accident both of his arms were crushed. There was no medical man within the distance of two hundred miles, and Captain Trevithick, believing that death would ensue if amputation was not immediately performed, offered his services, which were accepted by the patient. The operation, he informed me, was successful; the man rapidly recovered, and showed a pair of stumps which could have hardly been distinguished from the result of an operation by a regular surgeon. It is not improbable that in the warfare in which he had been engaged Captain Trevithick had been present and assisted at amputations of limbs of wounded soldiers. He thus probably acquired sufficient confidence to undertake and perform the operation himself.

" From Costa Rica Captain Trevithick came to England, with a design, among others, of forming a company to work a mine which had been granted him (for a term of years) by the Costa Rica Government. Mr. Gerard came to England with a similar object in view. Both failed in their object. Mr. Gerard was extremely unfortunate with regard to his mine, for he spent a considerable fortune of his own in working his mine to a loss.

" The eminence of Captain Trevithick as an engineer is well known. The public are indebted to him for the invention of the high-pressure steam-engine and the first railway steam-carriage. The latter being dependent on the former, Captain Trevithick informed me that the idea of the high-pressure engine occurred to him suddenly one day whilst at breakfast, and that before dinner-time he had the drawing complete, on which the first steam-carriage was constructed. Captain Trevithick informed me that in 1830 the original steam railway-engine constructed by him in 1808[1] at that time was still running in Wales."

---

[1] Probably referring to the Welsh locomotive of 1804.

"SIR,   "STANWIK, CUMBERLAND, 27th *November*, 1864.

"I read in the public prints that in a speech made by
you in Belle Vue Gardens you referred to the meeting of
Robert Stephenson with Trevithick at Carthagena, which, if
your speech be correctly reported, you attribute to accident.
The meeting was not an accident, although an accident led to
it, and that accident nearly cost Mr. Trevithick his life; and he
was taken to Carthagena by the gentleman that saved him,
that he might be restored. When Mr. Stephenson saw him he
was so recovering, and if he looked, as you say, in a sombre
and silent mood, it was not surprising, after being, as he said,
'half drowned and half hanged, and the rest devoured by
alligators,' which was too near the fact to be pleasant. Mr.
Trevithick had been upset at the mouth of the river Magdalena
by a black man he had in some way offended, and who capsized
the boat in revenge. An officer in the Venezuelan and the
Peruvian services was fortunately nigh the banks of the river,
shooting wild pigs. He heard Mr. Trevithick's cries for help,
and seeing a large alligator approaching him, shot him in the
eye, and then, as he had no boat, lassoed Mr. Trevithick, and
by his lasso drew him ashore much exhausted and all but dead.
After doing all he could to restore him, he took him on to
Carthagena, and thus it was he fell in with Mr. Stephenson,
who, like most Englishmen, was reserved, and took no notice of
Mr. Trevithick, until the officer said to him, meeting Mr.
Stephenson at the door, 'I suppose the old proverb of "two of
a trade cannot agree" is true, by the way you keep aloof from
your brother chip. It is not thus your father would have treated
that worthy man, and it is not creditable to your father's son
that he and you should be here day after day like two strange
cats in a garret; it would not sound well at home.' 'Who is
it?' said Mr. Stephenson. 'The inventor of the locomotive,
your father's friend and fellow-worker; his name is Trevithick,
you may have heard it,' the officer said; and then Mr. Stephenson
went up to Trevithick. That Mr. Trevithick felt the previous
neglect was clear. He had sat with Robert on his knee many a
night while talking to his father, and it was through him Robert
was made an engineer. My informant states that there was not

that cordiality between them he would have wished to see at Carthagena.

" The officer that rescued Mr. Trevithick is now living. I am sure he will confirm what I say, if needful. A letter will find him if addressed to No. 4, Earl Street, Carlisle, Cumberland.

" There are more details, but I cannot state them in a letter, and you might not wish to hear them if I could.

"I am, Sir,
" Your very obedient servant,
" JAMES FAIRBAIRN,
" who writes as well as rheumatic gout will let him.

" P.S.—I forgot to say the name of the officer is Hall.

" To — WATKIN, Esq."

"DEAR SIR,   "4, EARL STREET, CARLISLE, 16*th December*, 1864.

"On my return from Liverpool this day I find your letter of the 9th.

"In reply I have the honour to say that if you will be pleased to state upon what points you require information, I shall be but too happy to furnish it if I can.

"I have barely time to add that Mr. Fairbairn has left for America, which is his home, and has been for many years. He must have been at Birkenhead or Liverpool at the date of your letter to me. I was not aware that he had written to you. He brought me a paper with your remarks about the meeting of Mr. Robert Stephenson and Mr. Trevithick, and asked me if it were true that they met at Carthagena as stated, as he (Mr. Fairbairn) thought it was at Angostura, and that Mr. Trevithick was in danger of being drowned at the Bocasses, *i. e.* the mouths of the Orinoco, the Apure, &c., &c. I explained that it was near the mouth of the Magdalena.

"I will just say that it was quite possible Mr. R. Stephenson had forgotten Mr. Trevithick, but they must have seen each other many times. This was shown by Mr. Trevithick's exclamation, ' Is that Bobby?' and after a pause he added, ' I've nursed him many a time.'

" I know not the cause, but they were not so cordial as I could have wished. It might have been their difference of opinion about the construction of the proposed engine, or it might have been from another cause, which I should not like to refer to at present; indeed, there is not time.

" Pray address me as before. I hold no rank in the British service, and in England never assume any.

" I have the honour to be, dear Sir,

" Faithfully yours,

" BRUCE NAPIER HALL.
" EDWARD W. WATKIN, Esq., M.P., &c.,
" *Currente Calamo.*"

These notes from Mr. Hall and Mr. Fairbairn to Mr. (now Sir Edward) Watkin[1] arose from the latter repeating what Mr. Robert Stephenson had related of his meeting with Trevithick and Gerard at the inn at Carthagena. Stephenson said, " on his way home from Colombo, and in the public room at the inn, he was much struck by the appearance and manner of two tall persons speaking English; the taller of them, wearing a large-brimmed straw or whitish hat, paced restlessly from end to end of the room." Gerard and Stephenson entered into conversation, and Trevithick joined them. Stephenson said that he had a hundred pounds in his pocket, of which he gave fifty to Trevithick to enable him to reach England. It seems that had it not been for Mr. Hall's quick eye and steady hand rescuing Trevithick from the jaws of the blind alligator, he never would have returned to his native country.

Here was the inventor of the locomotive a beggar in a strange land, helped by the man whom he had nursed in baby-boyhood, then returning to England to

---

[1] Sir Edward Watkin contemplated writing a life of Trevithick.

become a great railway engineer in making known the use of the locomotive on the level road of the Liverpool and Manchester, while the real inventor, who looked upon railways and locomotives as things of a quarter of a century before, was about to recommend them as the means of passing across the isthmus of Costa Rica from the Atlantic to the Pacific, over the heights of the Cordillera, by the river San Juan from Greytown, and by its tributary the Serapique, then by railway towards the high ground of San José, the capital, and down the western slope, passing, somewhere not far from the mines, forward to the Gulf of Nicoya in the Pacific.

The approximate distance would be fifty miles of river navigation, and eighty or a hundred of railway, with perhaps stiffish but still manageable inclines, and no avalanches.

A loan by the Costa Rica Government of 1872 states,[1] "for the completion of the railway from the port of Limon, in the Atlantic Ocean, to San José de Costa Rica, and on to Heredia and Alajuela," near to Trevithick's mines, as if to carry out his design of forty-six years ago to connect the Atlantic and Pacific by railway.

_____

[1] See 'The Times,' 7th May, 1872.

CHAPTER XXIV.

RETURN TO ENGLAND.

IN the early part of October, 1827, the writer, then a boy at Bodmin school, was asked by the master if any particular news had come from home. Scarcely had the curiosity of the boys subsided, when a tall man with a broad-brimmed Leghorn hat on his head entered at the door, and after a quick glance at his whereabouts, marched towards the master's desk at the other end of the room. When about half-way, and opposite the writer's class, he stopped, took his hat off, and asked if his son Francis was there. Mr. Boar, who had watched his approach, rose at the removal of the hat, and replied in the affirmative. For a moment a breathless silence reigned in the school, while all eyes were turned on the gaunt sun-burnt visitor; and the blood, without a defined reason, caused the writer's heart to beat as though the unknown was his father, who eleven years before had carried him on his shoulder to the pier-head steps, and the boat going to the South Sea whaler.

During the next six months father and son sat together daily, the one drawing new schemes and calculations, the other observing, and learning, and calculating the weight and size and speed of a poor swallow he had shot, that the proportions of wings necessary to carry a man's weight might be known. In these calculations cube roots of quantities were extracted, which did not accurately agree with Trevithick's figures, who, asking for explanations, received a rehearsal, word for word, of the school-book rule for such extractions, which threw no more light on his understanding than did his own self-made rule on the writer's comprehension, though both methods produced nearly the same result.

Within a month of that time he heard of the arrival in England of Mr. Gerard, his companion in travel, from whom he had separated at Carthagena.

"My dear good Sir,   "Hayle Foundry, 15th November, 1827.

"I cannot express the extreme pleasure that the receipt of your favour of the 11th inst. from Liverpool gave me, as I had almost given up hopes of ever seeing you again, which you will see from the letters that I wrote Mr. Lowe; and after the severe rubs that we have undergone together, the parting us by shipwreck, as I supposed, at the close of our hardships, I doubly felt, and from your long absence, I supposed you must have encountered some severe gales; but thank God that we are safe landed to meet you and the dear boys again soon. We had a very good passage home, six days from Carthagena to Jamaica, and thirty-four days from thence for England; and on my return was so fortunate as to join all my family in good health, and also welcomed home by all the neighbourhood by ringing of bells, and entertained at the tables of the county and borough members, and all the first-class of gentlemen in the west of Cornwall, with a provision about to be made for me for the past services that this county has received from my inventions just before I left for Peru, which they acknowledge to be a saving in the mines since I left of above 500,000l., and that the present existence of the deep mines is owing to my inventions. I confess that this reception is gratifying, and have no doubt but that you will also feel a pleasure in it. I should be extremely happy to see you down here; it is but thirty-six hours' ride, and it will prepare you for meeting your London friends, as I would take you through our mines and introduce you to the first mining characters, which will give you new ideas and enable you to make out a prospectus that will show the great

advantages in Costa Rica mines over every other in South
America. I think it would not be amiss for you to bring with
you a few specimens, and after you have seen the Cornish mines
and miners I doubt not but we shall be able to state facts in so
clear a light that the first blow well aimed will be more than
half the battle, and prove a complete knock-down blow, which
in my opinion ought to be completed previous to your opening
your mining speculation in general in London. I have made a
very complete model of the gun, and it is approved of by all who
have seen it. Be so good as to remember me to the lads and
the Manilla man, and write me by return of post. I have not
as yet made any inquiry about the probability of getting adventurers for this new concern. I hope and trust that I shall
see you in Cornwall previous to our being together in London,
as it is my opinion that the nature of the concern requires it.

" I remain, Sir,

" Your humble servant,

(Signed) " RICHARD TREVITHICK.

" MR. JNO. GERARD,
" No. 42, St. Mary Axe, London."

Trevithick's hopeful character enabled him to enjoy
life in the midst of neglect and poverty. During the
eleven years of absence in America his wife and family
received no assistance from him. Shortly after leaving
his Quebrada-honda mountains of gold and silver, he
was penniless at Carthagena. On reaching England
he possessed nothing but the clothes he stood in, a gold
watch, a drawing compass, a magnetic compass, and a
pair of silver spurs. His passage-money being unpaid,
a chance friend enabled him to leave the ship. In a
month from that time he counted on getting a share of
the 500,000l. saved in the Cornish mines by the improvements he had effected in their steam-engines. The
ringing of bells and the talk of the neighbourhood
made him forget that he was a poor man, and the
Costa Rica mines were, he believed, soon to be in full

working, though not a single adventurer had been found.

The two lads Montelegre, coming to England to be educated, were sons of a gentleman of influence and authority in Costa Rica. On their perilous journey an attack of measles increased their discomforts. Probably one of those gentlemen has since filled the honourable position in this country of minister representing the Republic of Costa Rica.

"MY DEAR SIR,   "LONDON, *November* 17*th*, 1827.

"I arrived here from Liverpool last night, and this morning had the pleasure of receiving your kind letter of the 15th. The brig 'Bunker's Hill,' in which we came from Carthagena to New York, was wrecked within a few hours' sail of the port. We were in rather a disagreeable situation for some time, but more afraid than hurt. The cargo was nearly all lost. The ship was got off, but a complete wreck. The cause, however, of my delay in arriving arose from the want of the needful. You recollect Mr. Stephenson and Mr. Empson, agents for the Colombian Mining Association, whom we met at Carthagena. They kindly offered to supply me, but having determined to visit the celebrated Falls of Niagara, they insisted on my accompanying them, which I did.

"I am truly rejoiced to learn that your countrymen retain so lively a sense of the importance of your services. I think with you that before sounding the public or proceeding further, it might be well we should meet quietly to talk over everything and arrange our ideas, and that Cornwall, for the reasons you mention and others, would be the better place.

"The boys are well, and desire their respects to you.

"Your sincere friend,

"J. M. GERARD.

"CAPT. TREVITHICK."

Trevithick was friendly with George Stephenson when, in 1805, he nursed little Bobby. Twenty years afterwards, when George had comprehended Trevithick's locomotive, and desired his son's return to England to assist him in making it useful, Robert Stephenson, grown to manhood, met his father's friend in the wilds of Central America, both of them having been engaged in mining operations, and both on their return to England. George Stephenson's son made for himself a fortune and a name, his friend earned poverty and neglect. These two men, though well known to the engineering world, had no mutual attraction, and in their native land remained strangers to each other.

"My dear Sir,   "42, St. Mary Axe, *January 13th*, 1828.

"I had very unexpectedly a letter from Costa Rica this morning by the way of Jamaica, including two for you, which I have the pleasure of transmitting. Mine is from Montelegre, begun on the 25th of August, and finished on the 11th of September, when Don Antonio Pinto, with some people from the Alajuela, was to start by the road of Sarapique on his way to Jamaica. His intention was to find a better route as far as Buona Vista, after which he would probably nearly follow our course to the Embarcadero of Gamboa.

"Whether he succeeded in finding a less rugged road to Buona Vista I do not know. That he reached his destination seems clear from our letters having come to hand; but from their old date it would appear that he had either met with difficulties on the road or with considerable detention at San Juan. Montelegre writes me that Don Yonge had effected a compromise on your account with the Castros. Gamboa got back to San José on the 18th August, twelve days after he parted from us, to the great joy of our mutual friends. Mr. Paynter had been unwell after our departure. Both he and Montelegre desire their kindest recollections to you.

" Yours most sincerely,

" J. M. Gerard.

" Capt. Trevithick."

The newly discovered track taken by the homeward bound over the Cordilleras soon brought Don Antonio Pinto and others into the field in search of passable roads to the Atlantic. Twelve days required by Gamboa to effect his return to San José, a distance of perhaps sixty miles, indicate the difficulty.

Mr. Gerard passed some weeks with Trevithick in Cornwall arranging the best means of getting together a company to work on a large scale the Costa Rica mines.

"DEAR SIR, "HAYLE FOUNDRY, *January* 24*th*, 1828.

"Yesterday I saw Mr. M. Williams, who informed me that he should leave Cornwall for London on next Thursday week, and requested that I would accompany him. If you think it absolutely necessary that I should be in town at the same time, I would attend to everything that would promote the mining interest. When I met the Messrs. Williams on the mining concerns some time since, they mentioned the same as you now mention of sending some one out with me to inspect the mines, and that they would pay me my expenses and also satisfy me for my trouble with any sum that I would mention, because such proceedings would be satisfactory to all who might be connected in this concern. I objected to this proposal on the ground that a great deal of time would be lost and that the circumstances of your contracts in San José would not admit of such a detention; for that reason alone was my objection grounded, and if that objection could have been removed I should have been very glad to have the mines inspected by any able person chosen for that purpose, because it would not only take off the responsibility from us, but also strengthen our reports, as the mining prospects there will bear it out, and that far beyond our report. Some time since I informed you that I had drawn on the company for 100*l.* to pay 70*l* passage-money, and would have left 30*l.* to defray my expenses returning to London. The time for payment is up, but I have not as yet heard anything about it, therefore I expect there must be an omission by the bankers whose hands it was to have passed through for tendering it for payment. Perhaps in a day or two I shall hear something about it; I would thank you to inform me should you know anything about it. The unfavourable result of the gun I attribute in a great measure to the change in the Ministry and my not being present to explain the practicability of making the machinery about it simple. When Lord Cochrane has seen it, and a meeting takes place with him, my return to London may again revive its merits. This unfavourable report does not lessen its merits, neither will it deter me from again moving forward to convince the public of its practicability. I shall make immediately a portable model of the iron ship and engine, as they will be applicable to packets, which have been attempted at Falmouth, but found that the consumption of coals was so great that the whole of the ships' burthen would not contain sufficient coals to take them to Lisbon and return again, and on that account it was discontinued. That insurmountable object will now be totally removed, and I think that Lord Cochrane will make a very excellent tool to remove many weak objections made by persons not having sufficient ability to judge for themselves. His Lordship, being a complete master of science, is capable of appreciating their value from theory and from practice. I should not be surprised to see him down here to inspect it. It will be very agreeable if his Lordship comes here at the same time as yourself; he is a remarkably pleasant companion. My hearty thanks for your mother's good wishes towards me.

" Your humble servant,

" RD. TREVITHICK.

" MR. JNO. GERARD,
" No. 42, St. Mary Axe, London."

Gerard and Trevithick believed in the great value of the Costa Rica mines, and in the feasibility of working them profitably could capital sufficient be obtained. After a year or two passed in fruitless attempts to form a mining company in England, Mr. Gerard visited Holland and France with no better success; and while on this mission died in poverty in Paris, though brought up in youth as the expectant inheritor of family estates in Scotland. One of his letters says :—

"Robert Stephenson has given us his experience that it was unwise to take many English miners or workers to such countries. The chief reliance must after all be placed on the native inhabitants, under the direction and training of a small but well-selected party of Englishmen.

"Mining operations in that country are of such recent origin that a mining population can scarcely be said to exist. English workmen are not so manageable even in this country, and much less so in Spanish America, where they are apt to be spoiled by the simplicity and excessive indulgence even of the better classes, and where the high salaries they receive place them far above the country people of the same condition. All this tends to presumption and intolerance on their part, and ultimately to disputes and irreconcilable disgusts between them and the natives."

Mr. Michael Williams, Mr. Gibson, Mr. Macqueen, and others, were anxious to take up the mining scheme. The former proposed to send a person to examine the mines. This was a safe course, but not convenient to those who had made engagements to return without loss of time with miners and material to Costa Rica.

Mr. M. Williams informed the writer's brother that at a meeting of several gentlemen in London, a cheque for 8000l. was offered to Trevithick for his mining grant of the copper mountain in South America Words waxed warm, and the proffered money was refused. The next day Mr. Williams said to him, "Why did you not pocket the cheque before you quarrelled with them?" Trevithick replied, "I would rather kick them down stairs!"

In the end Trevithick got nothing for either his South American mines or those in Costa Rica.

CHAPTER XXV.

GUN-CARRIAGE — IRON SHIPS — HYDRAULIC CRANE — ICE MAKING —
DRAINAGE OF HOLLAND — CHAIN-PUMP — OPEN-TOP CYLINDER —
HAYLE HARBOUR — PATENT RIGHTS — PETITION TO PARLIAMENT.

"RICHARD TREVITHICK, of the parish of Saint Erth, in the
county of Cornwall, civil engineer, maketh oath and saith
that he hath invented new methods for centering ordnance on
pivots, fac litating the discharge of the same, and reducing
manual labour in time of action. That he is the true inventor
thereof, and that the same hath not been practised by any other
person or persons whomsoever to his knowledge or belief.

"Sworn, 10th November, 1827, before me, Rd. Edmonds."

"This gun is worked by machinery balanced on pivots giving
it universal motion, by one man, with the facility of a soldier's
musket. On one side a man puts in a copper charge of powder;
on the opposite side a man drops a ball in a bag down the gun,
as it stands muzzle up. The gunner, who sits on the seat
behind the gun, points it and pulls the trigger. The firing
causes it to run up an inclined plane at an angle of 25° for the
purpose of breaking the recoil; it runs down again with its
muzzle at the port, requiring no wadding, swabbing, cartridge,
or ramming, but runs in, out, primes, cocks, shuts the pan, and
breaks the recoil of itself; and by three men can be fired three
times in a minute with accuracy. The gun-carriage is a tube
3 feet long and 3 feet diameter, made of wrought-iron plate
$\frac{1}{4}$ of an inch thick, centered on a pivot to the deck, with the
gunner's seat attached, from which he looks through the case.
As the gun requires no tackle, and but a man on each side to
work it, only a space of 5 feet 6 inches is required from centre
to centre of ports, therefore a single-deck ship will carry a
greater number of guns than are now carried on a double-deck

ship, be worked with one-third of the hands, and be fired five times as fast as at present. A frigate would mount fifty

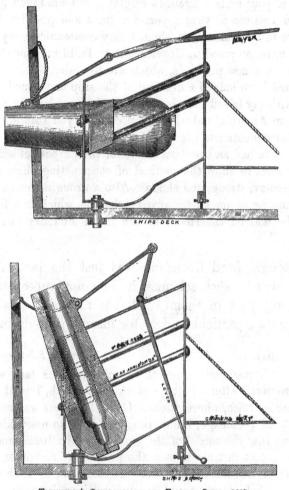

TREVITHICK'S GUN-CARRIAGE AND FRICTION SLIDES, 1827.

42-pound guns on one deck, with 150 men, and would discharge in the same time a greater weight of ball with greater precision than five 74-gun ships."[1]

---

[1] Description in Trevithick's handwriting.

"HAYLE, CORNWALL, 21st *February*, 1828.

" MY LORD COCHRANE,[1]

   " With great pleasure I read in the papers the announcement of your arrival again in England, and am much gratified
to find a person of your superior natural and practical talents,
so rare to be obtained, to whom I may communicate my views.

   " I have proposed to Government to build an iron ship, and
a gun on a new principle, which are to undergo an investigation, and have lodged a drawing of the ship and a model of the
gun with my friend Mr. Gerard, a gentleman who returned with
me from America, and who will present to you this letter with
the above-mentioned drawing and model.

   " I have had an iron boat made for the purpose of sending it
to London, to show the method of constructing ships on this
plan, roomy, strong, and cheap.   Also a wrought-iron ship with
a steam-engine on an improved principle, which in a few days
will be laid on the stocks at the Hayle Foundry iron manufactory."

Though Lord Cochrane was just the person to be
interested in such schemes, it does not appear that he
took any part in them.   At that time he was at work
on his own particular ideas for marine propulsion.

" MY DEAR SIR,   "LONDON, *February*, 1828.

   "Immediately after the receipt of your last, which I
only received after twelve o'clock on the 7th, I went to the
Ordnance Office, where, though Colonel Gossett was no longer
an official personage, I had the good luck to meet him.   He
told me that the model of the gun was at Woolwich, and could
not be got at in time to stop the progress of the other patent,
and which he considered of but little moment, as he thought it
very unlikely there could be any collision between the two
inventions.   He likewise said that from the official changes
that had taken place in the office, much loss of time might be
incurred by recalling the model, which was in train of being
examined.   To-day I have received a letter addressed to you

---

[1] Rough draft, by Trevithick, of unfinished letter.

from the Ordnance, by which it appears that your model has passed through an unsuccessful ordeal before the special committee.

"'SIR,   "'OFFICE OF ORDNANCE, 21st *February*, 1828.

"'I am directed by the Master General to acquaint you that the Select Committee of Artillery Officers, to whom your model of a 42-pounder carronade and carriage on a new principle were referred, have reported that on examination of the invention, they consider it to be wholly inapplicable to practical purposes. Your model is at the Ordnance Office, and will be delivered on your sending for it.

"'I am, Sir,

"'Your most obedient humble servant,

"'LOWNDES.

"'R. TREVITHICK, Esq.'

"My poor mother, who I regret to say has been very delicate ever since your departure, and is now again confined to bed, desires me to say that she is very sorry she is not Master General of the Ordnance, to give it a fair *practical* trial, as she thinks Captain Trevithick's opinions, though she cannot pronounce his name, may be fairly placed in opposition to that of the special committee of artillery officers.

"Ever faithfully yours,

"J. M. GERARD."

The recoil gun-carriage was his first occupation after twelve years of travel in countries where mechanical appliances were less thought of than weapons of war. He commenced this, his second era of inventions, with what he called a new thing, though it was but an extension of his schemes of 1809, when be patented iron vessels, hollow sliding masts and yards, self-reefing sails, and sliding keels.

The model gun was of brass, resting on a railway formed of two inclined bars of iron, up which the recoil propelled it into a convenient position for cleaning and loading. Its own gravity caused it to fall into the required place for being again fired. The slides also served as friction-bars to regulate the recoil.

The gun and the slides carrying it were enclosed in a wrought-iron box, having openings in the front and rear for the passage of the muzzle and the breech. The muzzle front of the box was pivoted to the deck by a strong bolt as a centre of motion, whilst its rear was supported on two small wheels resting on the deck, allowing the gun to change its line of horizontal fire by sweeping from the centre pivot. The gunner's seat moved with the carriage, from which he could elevate or depress the muzzle by a lever. The gun was self-priming and self-cocking; the powder charge was enclosed in a copper case. Captain Moncrieff's patent gun-carriage of the present day is described in words somewhat like those used by Trevithick forty years before. "The recoil lifted a weight smoothly and without friction; the gun and the weight were held in the position arrived at by a catch until the gun was loaded and ready to fire again."[1]

The iron boat mentioned in his note to Lord Cochrane as being made at Hayle, was "for the purpose of sending to London to show the method of constructing ships on this plan, roomy, strong, and cheap," and was thus spoken of in a newspaper of the 26th April, 1829. "The 'Scotsman' alludes to the intended construction of iron steamboats at Glasgow by Mr. Neilson:—" For fear of the public being misled on this subject, we beg to state that so far back as last Christmas twelvemonths we saw Trevithick, of Cornwall, superintending the construction of an iron man-of-war launch, with the

---

[1] See 'The Times,' August 12th, 1870.

avowed intention of applying a similar principle of construction to the building of fast-sailing iron steamboats." This intimation, in 1829, to the since famous Glasgow iron-ship builders, that they could not claim the invention because Trevithick had made such a boat in 1827, was probably in ignorance of Trevithick's patent and models of 1809,[1] explaining the advantages of ships of iron, either under sail or under steam, for commerce or for fighting-ships. The improved high-pressure steam-engine then in hand for iron ships was but the perfecting of his plans of twenty years before.[2]

"LAUDERDALE HOUSE, HIGHGATE, *April* 19*th*, 1830.

" MR. GILBERT,

" Sir,—I find by looking into the 'Art of Gunnery' that a 42-lb. shot discharged at the rate of 2000 feet a second in vacuum would send it to the height of 63,360 feet, which multiplied by the weight of the shot would be 2,661,120 lbs., with 12 lbs. of powder; and as guns, after being heated to about the heat of boiling water, will recoil their usual distance with half their first charge of powder, it proves that one-half the powder at first is lost in heating the gun to about 212°, which is a great deal under the heat of fired powder, therefore only 6 lbs. of powder effective force is applied to the ball. Now suppose this 6 lbs. of powder to be one quarter part carbon, $1\frac{1}{2}$ lb. is all the heat that can possibly be applied to perform this duty; then 1 lb. of carbon would be equal to 1,774,080 lbs. of duty actually performed; but if you take into calculation the great loss of power by the powder not being instantly all set on fire, with the gun so much below the heat of fired powder, the windage by the sides of the shot, the ball flying from the powder, and the immense power remaining in the gun at the time of the ball leaving its muzzle; if this was applied expansively, as in a cylinder, it may fairly be said to have double this power, or 3,548,160 lbs. for 1 lb. of carbon consumed, which, multiplied by 84, being the pounds in 1 bushel

---

[1] See vol. i., p. 302. [2] See vol. i., p. 329.

of carbon, gives 300 millions of duty. If it was applied to the best advantage, say on a piston, calling powder one thousand atmospheres, it would far exceed that duty. A gun 9 feet long and 7-inch bore has 16 feet of cold sides, and condenses at first one-half of its force by its cold sides and loses 150 millions in a 200th part of a second, while the ball passes from the breech to the muzzle. This gives 221,760 lbs. condensed by each foot of surface sides in so short a time. Binner Downs cylinder was taken as condensing 2500 lbs. for each surface foot in six seconds; therefore, without taking into account the great difference in time, there is eighty-eight times as much power lost by each foot of cold sides of the gun as by the cylinder sides. This shows what a considerable power is lost by cold sides where the vapour is so rare. Boulton and Watt's engine, doing twenty millions, performs with 1 lb. of coal a duty of 240,000 lbs., or about $\frac{1}{14}$th part of what is done by 1 lb. of carbon in powder. The water evaporated by the boiler is 7 lbs. thrown into steam by 1 lb. of coal, and a duty of 33,750 lbs. for each pound of water evaporated.

"Suppose 1 lb. of powder to contain 12 oz. of nitre and 4 oz. of carbon, and $\frac{1}{24}$th part of the nitre to be a fixed water, which would be half an ounce of water in every pound of powder, making the carbon eight times as much as the water; from this data 1 lb. of water in powder would perform a duty of 28,385,280 lbs.

	lbs.	
1 lb. of carbon in powder	3,548,160	} 14 times the consumption by the engine.
1 lb. of coal in Boulton and Watt's engine	240,000	
1 foot of cold sides of the gun	221,760	{ 88 times as much loss by the cold sides of the gun.
1 „ „ of the cylinder	33,750	
1 lb. of coal for 7 lbs. of water in steam		{ 14 times as much coal for water into steam as for water in powder.
1 lb. of carbon for 8 oz. of water in powder		

"By this it appears that heat is loaded with fourteen times as much water in steam-engines as in powder, and does only $\frac{1}{14}$th part of the duty of the water in powder. It is possible to heat steam independent of water, because if we work with steam of ten atmospheres, it would have ten times the capacity for heat, being in proportion to its gravity. The boiler standing on its end, with the fire in the bottom, and the water 1 foot thick above it, with a great number of small tubes from bottom to top, having great surface sides to heat the steam above the water, by working with a low chimney and slow fire, the tubes in the steam part of the boiler would not exceed 600° or 700° of heat, which would not injure them; as less water would be generated into steam, a very small part of the boiler would be sufficient for it; and as the coal required would be less, the boiler required would be very small. I state the foregoing to remind you that but little is yet known of what heat may be capable of performing; as this data so far exceeds whatever has been calculated on the power of heat before, when compared with steam in an engine.

" The power is sure, if we can find how to conduct it.

" I remain, Sir,

" Your very humble servant,

" RICHARD TREVITHICK.

" If you can spare time please to write to me."

The foregoing may be classed either under cannon or steam-engine; Trevithick combined them under the general laws of expansion by heat. Three years had passed since the committee of artillery officers sitting on his gun had given a verdict of no go; yet the subject was not forgotten, and his calculations enabled him to discover the explosive force, and the speed of the projectile in different parts of the gun, things which are now ascertained by mechanical tests and measures.

If a 7-inch cannon 9 feet long loses by absorption of heat during the time of the passage of the shot to the muzzle one-half of the expansive force of the powder, it is time to wrap our guns as well as our steam-engines in non-conductors. The greater heat of exploded powder than of steam caused eighty-eight times the amount of loss from abstracted heat, and yet the force from a pound of carbon in powder, was fourteen times as much as the Watt engine gave from a pound of coal.

"MR. GIDDY, "LONDON, No. 42, ST. MARY AXE, *June 18th*, 1828.

"Sir,—A few days since a Mr. Linthorn called on me and requested me to accompany him to Cable Street, near the Brunswick Theatre, to see a crane worked by the atmosphere, in a double-acting engine attached to it. He has a patent, and has entered into a contract with the St. Katharine's Dock Company to work their cranes, 140 in number, by a steam-engine of sufficient power to command the whole of them, by placing air-pipes around the docks, with a branch to each crane. To each crane is fixed a 10-inch cylinder, 20-inch stroke, double-acting. The atmosphere pressing on the piston like steam, the air is drawn from the pipes by a large air-pump and steam-engine.

"On being requested to give my opinion on this plan, after seeing one crane worked, I informed them of the disappointment that the ironmaster, Mr. Wilkinson, in Shropshire, several years since experienced, on the resistance of air in passing through long pipes from his blast-engine to his furnaces. He said he was aware of that circumstance, and it had since been further proved in London by one of the gas companies attempting to force gas a considerable distance, and who also failed.

"He thought that forcing an elastic fluid, and drawing it by a vacuum, were very different things, and that the error was removed by drawing in place of forcing. For my part I am not convinced on this head; but am still of opinion that the result on trial will be found nearly the same. However, let that be as it may, the expense and complication of the machine, having a double engine, with its gear attached to every separate crane, together with the immense quantity of air thrown into the air-pump from 140 double engines of 10 inches diameter, 20-inch stroke, eighty strokes per minute, and considering the numerous air leaks in such an extent of pipes and machines, must reduce the effect of the pressure of the atmosphere on each piston to a comparatively small power, unless the air-pump and steam-engine are beyond all reasonable bounds.

"Those objections I made them acquainted with, and said

that, before they went to such an expense, it would be a safer plan to first make further inquiry, so that their first experiment might be on a sure plan, for the other dock companies were looking for the results of this experiment.

" At the time I was informed of this plan, a thought struck me that it might be accomplished by another mode preferable to this: by a steam-engine to force water in pipes round the dock, to say 30 or 40 lbs. to the inch, more or less, and to have a worm-shaft, working in a worm-wheel, the same as a common roasting-jack, and apply to the worm-shaft a spouting arm like Barker's mill; the worm-shaft standing perpendicular would work the worm-wheel fixed in the chain-barrel shaft of the crane.

" This would make a very simple and cheap machine, and produce a circular motion at once, instead of a piston alternating motion to drive a rotary motion. My report had some weight with them; inquiry is to be made into the plan proposed by me, so as to remunerate me, provided my plan is considered good. Mr. Linthorn wishes an investigation before scientific and able judges, and requested me to name some one. I must again make free in asking the favour of your advice (which you have so ably given me for thirty years) on this plan. Mr. Linthorn intends to request Dr. Wollaston to accompany you, any day convenient to you. In the meantime, should you see him, it might not be amiss to mention it to him; and should you be able to attend for an hour or two to this business, I would thank you to drop me a note, saying when it may be convenient. There is a memorandum of an agreement between Mr. Linthorn and me; but the plan I suggest is only at present made public to him and yourself.

" Your most obedient servant,

" RICHARD TREVITHICK."

The reduction of friction by the use of an air-vacuum engine for working cranes, as designed by Mr. Linthorn, in lieu of an air-pressure engine, was doubted by Trevithick.

The Mont Cenis pneumatic-pressure machines which the writer saw at work lost much power by friction before experience had taught remedies. The pneumatic vacuum tubes which propelled the trains on the South Devon Railway, failed to give the power that was expected. Sir William Armstrong's hydraulic cranes, brought into use not many years after the date of Trevithick's letter, have been found effective. The writer, not knowing that Trevithick had before recommended hydraulic cranes for warehouses, accompanied Sir William over his works, then being erected near Newcastle-on-Tyne, and talked with him on the detail of his crane designs.

Trevithick thought of giving circular motion to the crane chain-barrel by the attachment of a screw-propeller, acted on by the force of a current of water at a pressure of 30 or 40 lbs. to the inch. Sir William Armstrong's arrangement was quite different; the merit due to Trevithick was for having pointed out the suitability of water as a means of conveying power through warehouses where fire was inadmissible.

" MR. GILBERT,   " LONDON, 42, ST. MARY AXE; *June 29th*, 1828.

"Sir,—Fancy and whim still prompt me to trouble you, and perhaps may continue to do until I exhaust your patience. A few days since I was in company where a person said that 100,000*l.* a year was paid for ice, the greatest part of which was brought by ships sent on purpose to the Greenland seas. A thought struck me at the moment that artificial cold might be made very cheap by the power of steam-engines; by compressing air in a condenser surrounded by water, and an injection to the same, so as to instantly cool down the highly-compressed air to the temperature of the surrounding air, and then admitting it to escape into liquid. This would reduce the temperature to any state of cold required.

"I remain, Sir,

"Your very humble servant,

"RICHARD TREVITHICK."

Trevithick's ideas for making ice have since been patented and made useful, though the detail of the operation has been improved by experience.

The Dutch, extending the use of steam on the Rhine and also in sea-going ships, wished Trevithick to see what was going on in Holland, where his nephew, Mr. Nicholas Harvey, was actively engaged in engineering. He had not money enough for the journey, and borrowed 2*l.* from a neighbour and relative, Mr. John Tyack. During his walk home a begging man said to him, "Please your honour, my pig is dead; help a poor man." Trevithick gave him 5*s.* out of the 40*s.* he had just begged for himself. How he managed to reach Holland his family never knew; but on his return he related the honour done him by the King at sundry interviews, and the kindness of men of influence in friendly communion and feasting.

" Mr. Gilbert,   "London, *July* 31*st*, 1828.

"Sir,—The night before last I arrived from Holland, where I spent ten days. I found my relative there, Mr. Nicholas Harvey, the son of John and Nancy Harvey. He is the engineer to the Steam Navigation Company at Rotterdam. They have a ship 235 feet long, 1500 tons burthen, with three 50-inch cylinders double, also two other vessels 150 feet long, each with two 50-inch cylinders double, ready to take troops to Batavia. The large ship with three engines cost 80,000*l.* The Steam Navigation Company built them, and many others of different sizes. This company has been anxious to get me to Holland, having heard of the duty performed by the Cornish engines. They were anxious to know what might be done towards draining and relieving Holland from its ruinous state.

"Immediately on arrival I joined the Dutch company, and entered into bonds with them.

"I give you, as near as I can, the present state of the country. About 250 years since, a strong wind threw a bank of sand

x 2

across the mouth of the river Rhine, which made it overflow its banks; 80,000 lives were lost, and about 40,000 acres of land, which remain to this time under 12 feet of water.

"About 100 years since the head and surface of the river Rhine was 5 feet below what it now is. The under floors of houses in Holland are nearly useless, and in another century must be totally lost, unless something is done to prevent it. The river at present is nearly overflowing its banks. In consequence of the rise of water, the windmill engines cannot lift it out. To erect steam-engines, they never could believe would repay the expense. Nearly one-half of Holland is at present under water, either totally or partially, because the ground kept dry in winter is flooded in summer.

"About six years since it was in contemplation to recover the 40,000 acres before mentioned, and a company was formed of the King and the principal men in Holland, to drain this by windmills, which they estimated would cost 250,000l., and making the banks and canals 450,000l. more, when made by men's labour, and seven years to accomplish it.

"This seven years was a great objection, because of the unhealthy state of the country while draining. The water is about 18 inches every year, to be lifted on an average 10 feet high. I have been furnished with correct calculations and drawings from this company.

"They expected to have drained 40,000 acres in seven years, at a cost of 700,000l., which, when drained, would have sold at 50l. per acre, about two millions.

"I find, from the statement given me, of 18 inches of water to be lifted 10 feet high, it would require about one bushel of coal to lift the water from one acre of ground for one year, and that a 63-inch cylinder double would perform the work of 40,000 acres, when working with high steam and condensing, at an expense of less than 3000l. per year. Engines in boats would cut and make the embankments and canals, without the help of men. I proposed six cylinders of 60 inches diameter, double power, which would drain the water in one year; and also four others for cutting the canals and making embankments. The expense would not exceed 100,000l. and one year, instead of

700,000*l.* and seven years. Above 60,000 acres more are to be drained.

"It was also proposed by Government to cut open the river Rhine to 1000 yards wide and 6 feet deep for 50 or 60 miles in length; they supposed it would cost them ten millions sterling. I proposed to make iron ships of 1000 tons burthen, with an engine in each, which would load them, propel, and also empty them for about 1*d.* per ton. Each ton will be about a square yard, and the cutting the river Rhine 1000 yards wide, 6 feet deep, 50 or 60 miles in length, will not cost one and a half million, and be accomplished in a short time. I further proposed that all this rubbish be carried into the sea of the Zuyder Zee, which would make dry, by embanking with the rubbish, nearly 1,000,000 acres of good land, capable of paying ten times the sum of cutting open the river Rhine.

"All this would add 100 per cent. more to the surface of Holland, and at this time it is much wanted, because their settlements abroad are free almost of the mother-country, and they have too many inhabitants for the land at present. I made them plans for carrying the whole into effect, and have closed my agreement with them.

"In a few days I shall go to Cornwall, and promised to return again to Holland within a month. I saw Mr. Hall and the engineer of the Dock Company to-day. They are satisfied that the plan for working the cranes is a good one. I am to see them again on Monday next; after which I shall return home, where I hope to see you, to consult you on the best plan for constructing the machines for lifting the water, cutting the canals, and making the dykes.

"I remain, Sir,

"Your very obedient servant,

"RICHARD TREVITHICK."

In this mere outline of a life it is impossible to go fully into the merit of Trevithick's plans for doubling the land surface of Holland. A drainage company was formed in London with a board of directors, some of whom thought that a new kind of engine should be invented and patented as a means of excluding others from carrying on similar but competing operations. Trevithick, always ready to invent new things, though never forgetting his experience with old things, instinctively returned to the Dolcoath engines, and recommended them as suitable for the pumping work; but finally a new design was determined on, and Harvey and Co., of Hayle, received orders for the construction, with the greatest possible dispatch, of a pumping engine for Holland.

This happening shortly after the writer had been taken from the Bodmin school, he was desired to help in the erection of this engine, and after working-hours made a drawing of its original form.

Plate XV. *a*, iron barge; *b*, wood frame supporting pump; *c*, open-top steam-cylinder 3 feet diameter, 8-feet stroke; *d*, piston guide-wheel; *e*, connecting rod; *f*, fly-wheel; *g*, cranked axle working air-pump bucket; *h*, connecting rod for air-pump bucket; *i*, air-pump; *j*, condenser; *k*, steam and exhaust nozzles; *l*, eccentrics working steam and exhaust valves; *m*, steam-pipe; *n*, cylindrical boiler, with internal fire-tube; *o*, external brick flues; *p*, chimney; *q*, feed-pump; *r*, feed-pipe; *s*, cup or rag-wheel; *t*, rag-chain, with iron balls; *u*, pump-barrel, 3 feet diameter; *v*, wheel guiding balls into bottom of pump-barrel; *w*, launder.

After a few successful though noisy trials, an alteration was made in the endless chain and in the guide-roller near the pump bottom. An amount of slack in the chain caused the balls to knock on passing this roller before entering the pump bottom. A chain having long links or bars of iron of uniform length, from ball

PLATE 15.

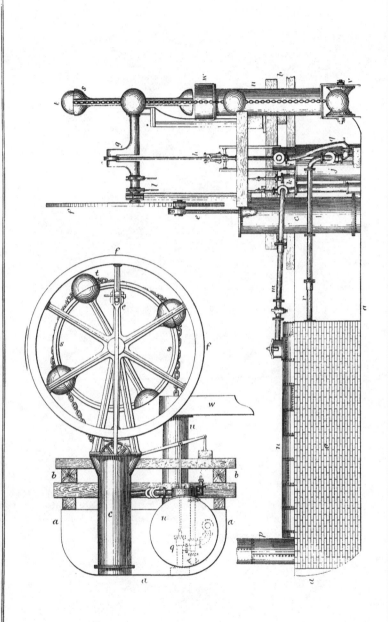

TREVITHICK'S CHAIN AND BALL PUMP

London: E. & F.N. Spon, 48, Charing Cross   Kell Bros. Lith London

to ball, jointed together by cross-pins, was substituted for the short link chain, and passed over a revolving hollow square frame at the bottom of the pump, in place of the curved roller-guide in the drawing. Each of the four sides of this square hollow frame was of the same length as the jointed link, and the balls lay in the hollow of the frame without touching it, contact being only on the links. The balls were thus guided directly into the bottom of the pump on their upward course with a rigid chain, and the swing and knocking was avoided. This pump was in principle the traditional rag-and-chain pump of a hundred years before; yet no trace of its use is met with during Trevithick's life in Cornwall. The early pump had rag balls, in keeping with the mechanical ignorance of the time, and suitable to man's power.

Trevithick's pump with iron balls raised "7200 gallons of water 10 feet high in a minute with 1½ lb. of coal,"[1] retaining all the original simplicity of the earlier rag-pump, having uniform circular motion and constant stream, without the use of a single valve. The engine and pump are thus described by him:—

"The first engine that will be finished here for Holland will be a 36-inch cylinder and a 36-inch water-pump, to lift water about 8 feet high. On the crank-shaft there is a rag-head of 8 feet diameter, going 8 feet per second, with balls of 3 feet diameter passing through the water-pump, which will lift about 100 tons of water per minute. It is in an iron boat, 14 feet wide, 25 feet long, 6 feet high, so as to be portable and pass from one spot to another without loss of time. This will drain 18 inches deep of water (the annual produce on the surface of each acre of land) in about twenty minutes; to drain each acre with about a bushel of coal costing 6d. per year. The engine is high pressure and condensing."[2]

---

[1] See letter, vol. ii., p. 332.   [2] See Trevithick's letter, vol. ii., p. 315.

It was something like the Newcomen open-topped cylinder of a hundred years before, but with a heavy piston, on the top of which a guide-wheel equal in diameter to the cylinder turned on a pin, to which the main connecting rod was jointed. The guide-wheel prevented any tendency to twist the piston from the angular positions of the connecting rod, and allowed the crank-shaft to be brought comparatively near to the cylinder top. The boiler was cylindrical, of wrought iron, with internal fire-tube and external brick flues; and gave steam of about 40 lbs. on the inch above the atmosphere, which, acting under the piston, caused the up-stroke, an expansive valve reducing the average pressure in the cylinder by one-half. The down-stroke was made by the atmospheric pressure of 14 lbs. on the inch, on the piston, its lower side being in vacuum, together with the weight of the thick piston and connecting rod, and the momentum of the revolving parts.

My readers must not suppose that this was an attempt to revive the discarded Newcomen engine; the likeness was only apparent; its power was mainly from the use of strong expansive steam, giving motion in the up-stroke through a rigid connecting rod, with controlling and equalizing crank and fly-wheel. It was not, as the Newcomen,[1] dependent for its power on the atmospheric pressure; and having no cylinder cover, or parallel motion, or beam, was not a Watt engine, though it had the Watt air-pump and condenser.

The Dolcoath engines continued to work with open-topped cylinders a quarter of a century after the Watt patent; and when they had passed away, many of Trevithick's high-pressure steam-engines retained the same form of outline, but had neither cylinder covers,

---

[1] See vol. i., p. 5.

parallel motion, air-pump, nor vacuum. The agricultural engines of 1813[1] and the South American
engines of 1816[2] had neither cylinder cover nor any
other part of the Watt engine, yet they successfully
competed with it in power, economy, and usefulness.

This design reveals a stumbling-block that superficial
people fall over. The boiler in the boat was surrounded
by brick flues, while a life-long claim of Trevithick's is
that before his tubular boiler with internal fire, there
could not be a successful steamboat, because brick flues
were dangerous in sea-going vessels, but in an iron
boat in smooth water it answered its purpose without in
any respect falsifying Trevithick's former claims or plans.

The chain pumping machine was in an iron barge,
the 36-inch diameter pump fixed just outside the
bow, its lower end a foot in the water; its height of
8 or 9 feet enabled the water from the pump-head to
flow through launders over the banks of the lakes to be
drained. Some of the directors came to Hayle to see it
work, and were well pleased at the constant stream of
water rushing from the foaming pump-head into the
launders. The large size of the rag-wheel gave the
rapidly revolving chain and balls a great speed. In
passing through the pump each ball forced upwards the
water above it, and drew up after it the following
water; before any ball had passed out at the top of the
pump the following ball had entered its bottom. The
directors having desired the writer to take the engine
to Holland and set it to work with the least possible
delay, adjourned for refreshment before starting for
London. In those few minutes differences arose, resulting in the engine remaining for months in the barge,
and then going to the scrap heap.

---

[1] See vol. ii., p. 37. [2] See vol. ii., p. 208.

Years afterwards others acted on Trevithick's drainage ideas, and Harvey and Co. built Cornish pumping engines with steam-cylinders 112 inches in diameter, similar in principle to the Dolcoath engine[1] of 1816, which effectually drained the Haarlem lake.

The Rhine during 100 years, in its passage through the low flat lands, had by deposit raised the level of its waters 5 feet, threatening to overflow the embankments and drown the surrounding country, that to a large extent was at a lower level than the river. All drainage from such land had to be pumped over the river bank, in many places 10 feet above the cultivated surface. Windmills had been used as pumping power, and a company had contemplated laying out 700,000l. in windmills and canals for drainage.

If the surface water averaged 18 inches in depth yearly, Trevithick could by steam-engines drain an acre of land by the consumption of a bushel of coal yearly. Four engines with cylinder of 63 inches in diameter would drain 160,000 acres, and four smaller engines in barges with suitable apparatus were to cut canals and construct embankments. The deposit of a hundred years was also to be removed, and the Rhine deepened 6 feet for a breadth of 1000 yards, and a length of 50 or 60 miles, by steam-dredgers, as used twenty years before in deepening the Thames,[2] to be fixed in iron ships of a thousand tons burthen. The cost of dredging from the bed of the river into a barge would be 1d. per ton; but this would be more than repaid by making with it an embankment, enclosing the Zuyder Zee, which would then in its turn be drained and made pasture land.

Before leaving for America he had reported on the best means of improving St. Ives Bay.[3] Hayle Harbour

---

[1] See vol. ii., p. 168.   [2] See vol. i., p. 243.   [3] See vol. i., p. 343.

was a branch of it, and he now suggested to Mr. Henry Harvey methods for deepening and improving it. A rival company of merchants and engineers, known then as Sandys, Carne, and Vivian, after many fights had recourse to law on the question of the course of a stream which had been changed by alterations during the making of wharfs and channels for ships.

Trevithick made a model in wood, movable layers of which indicated changes of level caused by workmen at different periods, giving a different course to the river bed. Mr. Harvey's counsel, since known as Lord Abinger and Sir William Follett, complimented Trevithick on the facility of understanding the case by reference to the model. The writer having carried the surveying chain, was present at the trial at the Bodmin assizes in 1829.

" Mr. Gilbert, " Hayle Foundry, *September 14th*, 1829.

"Sir,—I expected to have seen you before this, but am detained by Mr. Harvey's attorney to settle the Foundry Quay. As I made the drawing and model of the disputed ground, and was examined in evidence in court, it was thought proper that I should be present at the time that Mr. Peters came to determine the boundary line between the two companies. This cannot be concluded for ten days.

"As I have been so long detained I wish to await your arrival in Cornwall for the purpose of trying the new engine while you are down, and will thank you to inform me when you intend to be with us.

"I remain, Sir,

"Your very humble servant,

"Richard Trevithick."

Erskine, who had expressed the opinion favourable to Trevithick's engine more than twenty years before,[1]

---

[1] See vol. ii., p. 129.

was in this trial the counsel for the opposing side. The
verdict was in favour of Mr. Harvey, or Trevithick's side.

A former chapter[1] speaks of promises to pay certain
savings by the use of Trevithick's inventions prior to
his leaving for America. The United Mines refused to
continue the payment, and on Mrs. Trevithick's application to Mr. Davies Gilbert for advice he kindly wrote
to the Williamses, who managed those mines, and received the following reply:—

"DEAR SIR,   "SCORRIER HOUSE, *November* 14*th*, 1820.

". . . with regard to Mrs. Trevithick's claims for savings
on engines at the United Mines, there is much to be said.

"Before Mr. Trevithick went abroad he sold half the patent
right to William Sims, our engineer, who very strongly recommended that two of the engines at the United Mines should be
altered to what he considered his patent principle, but the
alterations proved very inferior to his expectations, and to this
circumstance I attribute much of the objections in question.
Mr. Henry Harvey has perhaps told you who the partners are
in the patent, and when you next come into this county I shall
be much pleased to wait on you at Tredrea that you may hear
the whole of the case; and though the United Mines adventurers are far from being a united body, I am very sure my
sons, who are their managers, are desirous to recommend what
appears to them right, and they will with myself be obliged for
your opinion after you have heard the whole matter on both sides.

"Dear Sir,

"Yours very sincerely,

"JNO. WILLIAMS.

"DAVIES GILBERT, Esq., M.P."

The opinion of Mr. Williams' elder son, Michael, has
been given.[2] Some of the family were quakers. No
further money payment for the saving of fuel followed
this carefully civil note, until Trevithick, on his return

---

[1] See vol. ii., p. 108.
[2] See letter from Mr. Michael Williams, vol. ii., p. 109.

from America, called at Scorrier House in a very threatening attitude on 31st October, 1827, when Mr. Williams, sen., said his reason for not continuing the payment was from his belief that the term of the patent had expired. Then came the following lawyer's letter :—

" SIR,   " PENZANCE, 7th November, 1827.

" I was at Captain Trevithick's yesterday, who observed to me he saw you at Scorrier a few days ago, and requested you would be good enough to settle the arrears on the savings on some of the engines in the mines for which you acted, none having been paid for a year or two, when you stated that the payment had been discontinued on account of the patent having expired. I find on a reference to the patent that it will not expire till May, 1830.

" I am, Sir,

" Your obedient servant,

" RD. EDMONDS.

" JOHN WILLIAMS, Esq., *Scorrier*."

" SIR,   " HAYLE, *January 24th*, 1828.

" Yesterday I called on Mr. Williams, and after a long dispute brought the old man to agree to pay me 150*l.* on giving him an indemnification in full from all demands on Treskerby and Wheal Chance Mines in future. He requested that you should make out this indemnification. I could not possibly get them to pay more, and thought it most prudent to accept their offer rather than risk a lawsuit with them.

" I remain, Sir,

" Your obedient servant,

" RICHARD TREVITHICK.

" RD. EDMONDS, Esq."

" Treskerby and Wheal Chance were, I believe, the only mines that paid for the use of the pole patent. Mr. John Williams, sen., of Scorrier, was purser of those mines. The agreement was that patentees should have one-fourth part of the savings of coal above twenty-six millions. The one-half of this fourth part from these two mines for some years was about 150*l.* per annum. This did not relate to the boilers;

Trevithick unfortunately did not take out a patent for that improvement. The adventurers of two or three mines only had
the honesty to pay 100*l.* for each mine; others made use of it
without acknowledgment.

" RD. EDMONDS.

"PENZANCE, 12*th January*, 1853."

Such were the recollections of the family solicitor
many years after the events had passed. The cylindrical high-pressure steam boiler and engine was really
included in the patent of 1802; but frequent detail
changes, consequent on size and position and local
requirements, were made up to 1811, when a perfected
form was arrived at, which is still in use. In principle
it was unaltered and not materially different in form,
but being used for larger engines, looked different. The
inventor saw nothing in this difference, but the public
did, and in the absence of the only man who could
prove their error refused to pay on the plea of its not
being patented. On his return from America he demanded 1000*l.* from each of the large Cornish mines, as
a settlement in full for all benefits derived from the use
of the Trevithick high-pressure steam-boiler. He had
proved the weakness of the law years before, when three
eminent counsel had given opinions on the 1802 patent,
one of them believing the patent good, because the
principle contained was new; two of them feared that
similarity of details might invalidate it:[1] so he determined to apply to the Government for remuneration for
benefits that might be called national.

" MR. RD. EDMONDS,   "HAYLE, *December* 20*th*, 1827.

    "Sir,—I send the principal heads of what you will have
to put in form to lay before the House. It is very defective;
but you will be assisted by Captain Andrew Vivian, who can

---

[1] See vol. ii , p. 129.

give dates and particulars, having been engaged with Mr. Gilbert and Captain Matthew Moyle in making out the duty performed at that time by Boulton and Watt and Hornblower's engines. He can also give you the results of the late improvements, with much more information than I can give. I saw him yesterday for this purpose; he will assist you with all his power, and will call on you at Penzance on Friday or Saturday. As I shall with pleasure pay him for his trouble, you need not fear calling on him for what assistance you need.

"Mr. Gerard and I propose to leave this for London on Saturday. If you think it necessary to see me, let Captain Vivian know it, and all meet at my house. I have sent you one of the monthly reports, in which you will see John Lean's report of Dolcoath engines, from which I have given you in my statement the average results and savings.

"I remain, Sir,

"Your very humble servant,

"RICHARD TREVITHICK.

"P.S.—I was at Dolcoath account on Monday, and made known to them my intention of applying to Government, and not to individuals, for remuneration. They are ready to put their signatures to the petition, and so will all the county. I fear that it is as much as we shall do to get it before the House in time."

The following petition was drawn up and put into the hands of his old friend Davies Gilbert, then a Member of Parliament :—

"TO THE HONOURABLE THE COMMONS OF THE KINGDOM OF GREAT BRITAIN AND IRELAND IN PARLIAMENT ASSEMBLED.

"The Humble Petition of Richard Trevithick, of the Parish of Saint Erth, in the County of Cornwall, Civil Engineer, 27th February, 1828,

"SHEWETH:

"That this kingdom is indebted to your petitioner for some of the most important improvements that have been

made in the steam-engine, for which your petitioner has not
hitherto been remunerated, and for which he has no prospect
of being ever remunerated except through the assistance of
your Honourable House.

"That the duty performed by Messrs. Boulton and Watt's
improved steam-engines in 1798, as appears by a statement
made by Davies Gilbert, Esq., and other gentlemen associated
for that purpose, averaged only fourteen millions and half
(pounds of water lifted 1 foot high by 1 bushel of coals),
although a chosen engine of theirs, under the most favourable
circumstances, at Herland Mine lifted twenty-seven millions,[1]
which was the greatest duty ever performed till your petitioner's
improvements were adopted, since which the greatest duty
has been sixty-seven millions, being more than double the
former duty. That prior to the invention of your petitioner's
boiler the most striking defect observable in every steam-engine
was in the form of the boiler, which in shape resembled a tilted
waggon, the fire applied under it, and the whole surrounded
with mason-work. That such shaped boilers were incapable of
supporting steam of a high pressure, and did not admit so much
of the water to the action of the fire as your petitioner's boiler
does, and were also in other respects attended with many disadvantages.

"That your petitioner, who had been for many years employed in making steam-engines on the principle of Boulton
and Watt, and had made considerable improvements in their
machinery, directed his attention principally to the invention
of a boiler which should be free from these disadvantages; and
after having devoted much of his time and spent nearly all his
property in the attainment of this object, he at length succeeded
in inventing and perfecting that which has since been generally
adopted throughout the kingdom.

"That your petitioner's invention consists principally in
introducing the fire into the midst of the boiler, and in making
the boiler of a cylindrical form, which is the form best adapted
for sustaining the pressure of high steam.

"That the following very important advantages are derived

---

[1] See Mr. Taylor's report on Herland engine, vol. ii. p. 118.

from this, your petitioner's, invention. This boiler does not require half of the materials, nor does it occupy half the space required for any other boiler. No mason-work is necessary to encircle the boiler. Accidents by fire can never occur, as the fire is entirely surrounded by water, and greater duty can be performed by an engine with this boiler, with less than half the fuel, than has ever been accomplished by any engine without it. These great advantages render this small and portable boiler not only superior to all others used in mining and manufacturing, but likewise is the only one which can be used with success in steam-vessels or steam-engine carriages. The boilers in use prior to your petitioner's invention could never with any degree of safety or convenience be used for steam navigation, because they required a protection of brick and mason work around them, to confine the fire by which they were encircled, and it would have been impossible, independent of the great additional bulk and weight, that boilers thus constructed could withstand the rolling of vessels in heavy seas; and notwithstanding every precaution the danger of the fire bursting through the brick and mason work could never be effectually guarded against.

"That had it not been for this, your petitioner's, invention, those vast improvements which have been made in the use of steam could not have taken place, inasmuch as none of the old boilers could have withstood a pressure of above 6 lbs. to the inch, much less a pressure of 60 lbs. to the inch, or even of above 150 lbs. to the inch when necessary.

"That as soon as your petitioner had brought his invention into general use in Cornwall, and had proved to the public its immense utility, he was obliged in 1816 to leave England for South America to superintend extensive silver mines in Peru, from whence he did not return until October last. That at the time of your petitioner's departure the old boilers were falling rapidly into disuse, and when he returned he found they had been generally replaced by those of his invention, and that the saving of coals occasioned thereby during that period amounted in Cornwall alone to above 500,000*l*.[1]

---

[1] See Lean's report, vol. ii., p. 175.

"That the engines in Cornwall, in which county the steam-engines used are more powerful than those used in any other part of the kingdom, have now your petitioner's improved boilers, and it appears from the monthly reports that these engines, which in 1798 averaged only fourteen and half millions now average three times that duty with the same quantity of coals, making a saving to Cornwall alone of 2,781,264 bushels of coals, or about 100,000l. per annum. And the engines at the Consolidated Mines in November, 1827, performed sixty-seven millions, being forty millions more than had been performed by Boulton and Watt's chosen engine at Herland, as before stated.

"That had it not been for your petitioner's invention, the greater number of the Cornish mines, which produce nearly 2,000,000l. per annum, must have been abandoned in consequence of the enormous expense attendant on the engines previously in use.

"That your petitioner has also invented the iron stowage water-tanks and iron buoys now in general use in His Majesty's navy, and with merchant's ships.

"That twenty years ago your petitioner likewise invented the steam-carriage, and carried it into general use on iron rail-roads.

"That your petitioner is the inventor of high-pressure steam-engines, and also of water-pressure engines now in general use.

"That his high-pressure steam-engines work without condensing water, an improvement essentially necessary to portable steam-engines, and where condensing water cannot be procured.

"That all the inventions above alluded to have proved of immense national utility, but your petitioner has not been reimbursed the money he has expended in perfecting his inventions. That your petitioner has a wife and large family who are not provided for.

"That Parliament granted to Messrs. Boulton and Watt, after the expiration of their patent for fourteen years, an extension of their privileges as patentees for an additional period, whereby they gained, as your petitioner has been informed, above 200,000l.

"That your petitioner therefore trusts that these his own

important inventions and improvements will not be suffered to go unrewarded by the English nation, particularly as he has hitherto received no compensation for the loss himself and his family have sustained by his having thus consumed his property for the public benefit.

"Your petitioner therefore most humbly prays that your Honourable House will be pleased to take his case into consideration, and to grant him such remuneration or relief as to your Honourable House shall seem meet.

"And your petitioner, as in duty bound, will ever pray, &c.

"RD. EDMONDS,

"<i>Solicitor, Penzance.</i>"

From the Patent Office to the House of Commons was, for a petitioner, as bad as out of the frying-pan into the fire. Trevithick solicited the support of Members of Parliament until tired of running after friends, and the petition became a dead letter, though the mining interests of Cornwall had in twelve years saved 500,000*l.* by his unrewarded inventions.

"LAUDERDALE HOUSE, HIGHGATE, *December 24th,* 1831.

"MR. GILBERT,

"Sir,—I find that Mr. Spring Rice cannot get the Lords of the Treasury to agree to remunerate or assist me in any way. He appeared to be much disappointed, and said that he would write to the Admiralty Board on Thursday last, recommending them to adopt this engine. As yet I have heard nothing respecting it, nor do I expect to during the holy days; but in the interim I wish to look out for some moneyed man to join in it, otherwise I fear I shall lose the whole. Can you assist in recommending anyone you know? I wish Mr. Thompson would come into it, he would be a good man. Can you furnish me with a copy of your report to Mr. Spring Rice, or something relating thereto? It would be a great assistance in getting some one to join.

"The sum required is small, and the risk is less; but the

prospect is great, beyond anything I ever knew offered on such easy terms. Waiting your reply,

"I remain, Sir,

"Your very humble servant,

"RICHARD TREVITHICK."

"DEAR TREVITHICK,   "EASTBOURNE, *December 26th,* 1831.

"I am sorry to find that you have not any prospect of assistance from Government. I have not any copy or memorandum of my letter to Mr. Spring Rice; but it was to the effect of first bearing testimony to the large share that you have had in almost all the improvements on Mr. Watt's engine, which have altogether about trebled its power; to your having made a travelling engine twenty-eight years ago; of your having invented the iron-tanks for carrying water on board ships, &c.

"I then went on to state that the great defect in all steam-engines seemed to be the loss, by condensation, of all the heat rendered latent in the conversion of water into steam; that high-pressure engines owed their advantages mainly to a reduction of the relative temperatures of this latent heat; that I had long wished to see the plan of a differential engine tried, in which the temperatures and consequently elasticities of the fluid might be varied on the opposite sides of the piston, without condensation; that the engine you have now constructed promised to effect that object; and that, in the event of its succeeding at all, although it might not be applicable to the drawing water out of mines, yet that for steam-vessels and for steam-carriages its obvious advantages would be of the greatest importance; and I ended by saying that although it was clearly impossible for me to ensure the success of any plan till it had been actually proved by experiment, yet judging theoretically, and also from the imperfect trial exhibited on the Thames, I thought it well worthy of being pursued. Your plan unquestionably must be to associate some one with you (as Mr. Watt did Mr. Boulton), and I certainly think it a very fair speculation for any such person as Mr. Boulton to undertake.

"It is impossible for me to point out any individual, as never

having had the slightest connection with trade or with manufacture in any part of my life, I am entirely unacquainted with
mercantile concerns. I- cannot, however, but conjecture that you
should make a fair and full estimate of what would be the
expense of making a decisive experiment on a scale sufficiently
large to remove all doubt; and that your proposal should be,
that anyone willing to incur that expense should, in the event
of success, be entitled to a certain share of your patent. On
such conditions some man of property may perhaps be found
who would undertake the risk; and if the experiment proves
successful, he will be sure to use every exertion afterwards for
his own sake. With every wish for your success,

"Believe me,

" Yours very sincerely and faithfully,

" DAVIES GILBERT."

The petition to Parliament for a national payment
for national gains, so hopefully taken up on his return
from America, when experience had proved the value
of his inventions, after four weary years of deferred
expectation, was consigned to the tomb of forgetfulness.
Compare the petition of 1828 with a modern report.

" Prior to the invention of your petitioner's boiler, the most
striking defect observable in every steam-engine was in the
form of the boiler which in shape resembled a tilted wagon;
your petitioner's invention consists principally in introducing
the fire into the midst of the boiler, and in making the boiler of
a cylindrical form, which is the form best adapted for sustaining
the pressure of high steam, and does not require half of the
materials, nor does it occupy half the space required for any
other boiler, and greater duty can be performed by an engine
with this boiler with less than half the fuel, than by any
engine without it, and is the only one that can be used with
success in steam-vessels, as none of the old boilers could have
withstood a pressure of above 6 lbs. on the inch, much less a
pressure of 60 lbs. or even of 150 lbs. when necessary."

A report of the Royal Mail Steam Packet Company in 1871 states, "by placing compound engines in the 'Tasmania,' they had reduced the consumption of coal to one-half the former quantity, doubled her capacity for freight, and increased her speed."[1] Presuming that the compound engines of the 'Tasmania' are like other engines known by that name, having high-pressure steam in a comparatively small cylinder from which it expands in a larger one, tubular boilers, surface condensers, and screw-propeller, the saving admitted in the 'Tasmania' is just what Trevithick's petition pointed out forty-three years ago—to lessen by one-half the weight, space, and fuel in marine steam-engines—his opinion being founded on the experience of a lifetime, for as early as 1804 he wrote on the question of compound engines, "I think one cylinder partly filled with steam would do equally as well as two cylinders;"[2] and again in 1816, describing expansion, "The engine is now working with 60 lbs. of steam, three-quarters of the stroke expansive, and ends with the steam rather under atmosphere strong;"[3] and in the same year worked the expansive compound engine at Treskerby.[4]

---

[1] See 'The Times,' October 26th, 1871, Half-yearly Report of the Chairman.
[2] See vol. ii., p. 134.   [3] See vol. ii., p. 91.   [4] See vol. ii., p. 104.

CHAPTER XXVI.

TUBULAR BOILER, SUPERHEATING STEAM, AND SURFACE CONDENSER.

" MR. GILBERT,   " HAYLE FOUNDRY, *December 14th*, 1828.

"Sir,—On my return from London five weeks since I was disappointed at not finding you in Cornwall. I have made inquiry into the duty performed by the best engines, and the circumstances they are under, from which it appears to me there is something which as yet has not been accounted for, particularly in Binner Downs engines. A statement was given to me by Captain Gregor, the chief agent and engineer of the mine, which appears so plain that I cannot doubt the facts, though they differ very widely from all former opinions. There are two engines, one of 42 inches diameter, the other of 70 inches diameter, 10-feet stroke.

"Formerly those engines worked without cylinder cases, when the 70-inch cylinder burnt $1\frac{1}{2}$ wey of coal, and performed a regular duty of forty-one millions; since that time brickwork has been placed round the cylinder and steam-pipes, leaving a narrow flue, which is heated by separate fires. These flues consume about 5 bushels of coal in twenty-four hours; the heat is not so great as to injure the packing, which stands good for thirteen weeks; the saving for several months past has increased the duty to sixty-three millions.

"Before the use of this flue 108 bushels of coal were consumed under the boiler, now only 67 bushels are needed, which with the 5 bushels in the flue gives 72 bushels. The coal burnt under the boiler gives a duty of sixty-six millions, or an expansion of 60 per cent. by the heat of 5 bushels of coal in the flues, and a duty of 1781 millions gained in twenty-four hours by 5 bushels of coal, which amounts to 350 millions gained by each of these 5 bushels. The 42-inch cylinder is as near as possible under the same circumstances, no other alterations have been made; and to prove this they left out the fires in the flues, and the engines fell back to their former duty, and the condensing water increased in the same proportion.

" The surface sides heated by this 5 bushels of coal is about 300 surface feet, the saving effected is 1781 millions, which is six millions saving for each foot of surface on the castings in the flues.   In Wheal Towan engine that did eighty-seven millions, the surface sides of the boiler was 1000 feet of fire-sides for every bushel of coal burnt in an hour, and the duty performed per minute from each foot of boiler fire-sides was 1500 lbs. 1 foot high. Now it appears that the heating of Binner Downs 300 surface feet gave a saving of 6000 lbs. per minute per surface foot; whereas the boiler sides only gave 1500 lbs. of duty per minute for each foot of boiler fire-sides.   Therefore the saving by heating the sides of the cylinder is equal to four times the duty done by each square foot of boiler sides; and further. it appears that the 300 feet, when not heated, though clothed round with brickwork, condensed or prevented from expanding the steam of 41 bushels of coals, which was eight times as much steam condensed as the 5 bushels of coal would raise.   Now if this be a report of facts, which I have no reason to doubt (but still I will be an eye-witness to it next week), there must be an unknown propensity in steam above atmosphere strong to a very sudden condensation, and *vice versâ*, to also a sudden expansion, by a small heat applied to the steam-sides; and if by heating steam, independent of water, such a rapid expansion takes place, certainly a rapid condensation must take place in the same ratio, which might be done at sea by cold sides to a great advantage, always working with fresh water.

" I shall have a small portable engine finished here next week, and will try to heat steam, independent of water, in small tubes of iron, on its passage from the boiler to the cylinder, and also try cold sides for condensing.

" If the above statement prove to be correct, almost anything might be done by steam, because then additional water would not be wanted for portable engines, but partially condensed and again returned into the boiler, without any fresh supply or the incumbrance of a great quantity; and boilers might be made with extensive fire-sides, both to heat water and steam, and yet be very light.

"It appears that this engine, when working without the heated flues round the cylinder and pipes, evaporated 20,000 gallons of water into steam, in twenty-four hours, more than when the flues were heated, and the increase of condensing water was in the same proportion. It is so unaccountable to me that I shall not be satisfied until I prove the fact, the result of which I will inform you, and shall be very glad to receive your remarks on the foregoing statement.

"The first engine that will be finished here for Holland will be a 36-inch cylinder, and a 36-inch water-pump, to lift water about 8 feet high; on the crank-shaft there is a rag-head of 8 feet diameter, going 8 feet per second, with balls of 3 feet diameter passing through the water-pump, which will lift about 100 tons of water per minute. It is in a boat of iron, 14 feet wide, 25 feet long, 6 feet high, so as to be portable, and pass from one spot to another, without loss of time. It will drain 18 inches deep of water (the annual produce on the surface of each acre of land) in about twenty minutes for the drainage of each acre, with one bushel or sixpennyworth of coal per year. The engine is high pressure and condensing.

"I remain, Sir,

"Your very humble servant,

"RICHARD TREVITHICK.

"P.S.—Woolf is making an apparatus to throw back from the bottom of the cylinder on to the top of the piston a fluid metal every stroke. He says he proved by an indicator that he raised 18,000 inches of steam from 1 inch of water, of 11 lbs. to the inch pressure on a vacuum, and that the reason why this engine did not do 300 millions, was because the steam passed by the sides of the piston. That an engine at the Consolidated Mines working 10 feet 2 inch stroke, going $\frac{7}{8}$ths expansive, beginning with steam of 20 lbs. to the inch above the atmosphere, and ending with 11 lbs. on a vacuum. I doubt this statement; however, there is some hidden theory as yet, because some engines perform double as much as others, under the same known circumstances, and I believe that nothing but practice will discover where this defect is, for, in my opinion, no statement of theory yet given is satisfactory why high-pressure engines so far exceed low-pressure engines. It is facts that prove it to be so, therefore all theory yet laid down must be defective."

At the date of this letter Trevithick had been rather more than a year in England, residing generally at Hayle, within half-a-dozen miles of Mount's Bay, from

MOUNT'S BAY. [W. J. Welch.]

which he had sailed for America; and after eleven years of wandering in countries where steam-engines were unknown, except those that he himself had constructed, was again on his return giving his whole thoughts to the idol of his life.

During that period scientific men in Europe thought and wrote much on the question of relative temperature, pressure, economy, and manageability of steam. Newcomen's great discovery a century before was the avoidance of the loss of heat by the cooling at each stroke of the exterior of the steam-vessel of Savery's engine by injecting cold water into the steam in the cylinder. After fifty years came the Watt improvement, still reducing the loss of heat by removing the cold injection-water from the steam-cylinder to a separate condenser.

The high-pressure steam-engine was perfect without injection-water, though when convenient its use was equally applicable as in the low-pressure engine. Trevithick, on his return to civilized life, read the views of Watt on steam, as given in 'Farey on the Steam-Engine.' On informing Davies Gilbert of his doubts of the accuracy of those views, and of his intention of testing them by comparison with the work performed by Cornish pumping engines, his friend, who had just published his 'Observations on the Steam-Engine,'[1] forwarded a copy, from which the following is an extract :—

"One bushel of coal, weighing 84 lbs., has been found to perform a duty of thirty, forty, and even fifty millions, augmenting with improvements, chiefly in the fire-place, which produce a more rapid combustion with consequently increased temperature, and a more complete absorption of the generated heat; in addition to expansive working, and to the use of steam raised considerably above atmospheric pressure."

---

[1] 'Observations on the Steam-Engine,' by Davies Gilbert, V.P.R.S., January 25th, 1827. See 'Philosophical Transactions.'

Those words gave the result of Trevithick's experience made known to his friend during twenty years of labour,[1] and yet by a seeming fatality his name is not found in his friend's book.

Sir John Rennie, who in youth had been employed under Boulton and Watt at Soho, and had risen to be a member of the Royal Society, came about that time into Cornwall, at the request of the Admiralty, to make examination into the work performed by Cornish pumping engines, and selected Wheal Towan engine on which to make special experiments.[2] The subject of Trevithick's note was therefore at that period, and still is, a matter of importance; and his practical treatment of the question is more instructive to young engineers than complex rules. Arthur Woolf was at the same time experimenting on steam at the Consolidated Mines, and finding the want of agreement between the rules of low-pressure and the practice of high-pressure engines, imputed the error to the escape of steam by the sides of the piston. Trevithick disbelieved this, " because some engines perform double as much as others, under the same known circumstances," and advocated the observance of general practice to prove why high-pressure engines were more economical than those of low-pressure. Captain Gregor had placed fire-flues around the steam cylinder and pipes, hoping thereby to exceed the duty of the Wheal Towan engine, whose boiler, cylinder, and steam-pipes were carefully clothed with a thick coating of sawdust or other non-conductor of heat, and lifted eighty-seven millions of pounds of water 1 foot high by the heat from a bushel of coal weighing 84 lbs. This was the greatest duty that had ever been recorded from a steam-engine. The Trevithick or Cornish boilers,

---

[1] See letter, vol. ii., p. 143.   [2] See vol. ii., p. 185.

similar to those in Dolcoath,[1] measured at the rate of 1000 superficial feet of heating surface for each bushel of coal burnt in an hour, and in round numbers gave a duty of 1500 lbs. lifted a foot high to each foot of boiler surface. In words not technical, the heat from 1 lb. of coal gave steam that raised 460 tons weight of water 1 foot high.

The cylinder of this engine used the Watt steam-jacket. The Binner Downs engine was doing not one-half this duty, namely, forty-one millions; when brick flues were built around the cylinder, cylinder cover, and steam-pipes, and one or two fire-places, fixed near the bottom of the cylinder, of a size to conveniently burn 5 bushels of coal in twenty-four hours, the heat from which circulated through those flues on its way to the chimney, and increased the duty of the engine by one-half, raising it to sixty-three millions; in other words, during twenty-four hours of working, 67 bushels of coal in the boiler, and 5 bushels in the cylinder flues, did the same work as 108 bushels in the boiler without the cylinder flues, causing a saving of fifty per cent. by their use. Another startling fact was the greater effect for each foot of heating surface in the steam-cylinder flues than in the boiler flues; the latter gave a power of 1500 lbs. raised 1 foot high by a bushel of coal, while the former gave 6000 lbs. of power from the same amount of coal and heating surface.

Here was a mystery that Trevithick would not believe until he had seen it with his own eyes: he searched for it for a year or two, and overlooking the fact that the more simply arranged engine of his once pupil, Captain Samuel Grose, was doing more duty than the superheating steam-engine at Binner Downs, he

---

[1] See drawing, vol. ii., p. 169.

worked at what seemed to be new facts, and converted them into a new engine.

We have traced how succeeding engineers tried to prevent loss of heat. Trevithick took the first bold step, and aiming at the same object, made the boiler the steam-jacket for the cylinder, and in his patent of 1802 went still further and protected the boiler from external cold, and thus describes it :—" The steam which escapes in this engine is made to circulate in the case round the boiler, where it prevents the external atmosphere from affecting the temperature of the included water, and affords by its partial condensation a supply for the boiler itself."[1] So that a quarter of a century before the date of those Binner Downs experiments he had patented an engine having neither cylinder nor boiler exposed to the cooling atmosphere. The flues around the Binner Downs cylinder were difficult of control. Trevithick says the piston packing had not been injured, showing that observers thought it would be, and even the cylinder was endangered, for the writer, who stoked those heating flues, recollects the fires burning very brightly in them. The ready transmission of heat through thin metal, used by Trevithick in 1802 for heating feed-water, and in the cellular bottom of the iron ship of 1808, serving as a surface condenser,[2] and his experience in 1812, that "the cold sides of the condenser are suffi-cient to work an engine a great many strokes without any injection,"[3] still followed up in 1828 by condensing steam without the use of injection-water, led to what is since known as Hall's surface condenser.

The following letter is in the handwriting of the

---

[1] See patent specification, vol. i., p. 132.
[2] See vol. i., p. 335.
[3] See Trevithick's letter, 7th December, 1812, vol. ii., p. 18.

present writer; it is the only one of Trevithick's numerous letters not written by himself:—

"MR. GILBERT,   "HAYLE, *December 30th*, 1828.

"Sir,—On the 28th inst. I received your printed report on steam, and have examined Farey's publication on sundry experiments made by Mr. Watt, which are very far from agreeing with the actual performance of the engines at Binner Downs. Mr. Watt says that steam at one atmosphere pressure expands 1700 times its own bulk as water at 212°, and that large engines ought to perform eighteen millions when loaded with 10 lbs. to the inch of actual work, the amount of condensing water being one-fortieth part of the content of the steam in the cylinder at one atmosphere strength, the cold condensing water at 50°, and when heated 100°. This would give for the Binner Downs engine, with a 70-inch cylinder, 10-inch stroke, 11 lbs. effective work on the inch (this load being one-tenth more than in Watt's table, by Farey, for an engine of this size and stroke), 57 gallons of injection-water for each stroke, and when working eight strokes per minute, to do eighteen millions would consume $11\frac{1}{4}$ bushels of coal per hour.

"Now the actual fact at Binner Downs, at the rate of working and power above mentioned, is that 3 bushels of coal per hour were burnt, using 13 gallons of injection-water at each stroke at 70° of heat, which was raised by its use to 104°, or an increase of 34°, which, multiplied by 13 gallons, gives 442. Mr. Watt's table for this engine and work gives 57 gallons of condensing water at 50°, heated by use to 100°. This 50° raised, multiplied by the 57 gallons of water, amounts to 2850, or six and a half times the quantity really used in the Binner Downs engine, and nearly four times the coal actually used at present. Mr. Watt further says that steam of 15 lbs. to the inch, or one atmosphere, from 1 inch of water at 212° occupies 1170 inches, and that steam of four atmospheres, or 60 lbs. to the inch, gives only 471 inches at a heat of 293°. Now deducting 50° from 212° leaves 162° of heat raised by the fire. Multiply 15 lbs. to the inch by 1700 inches of steam, and divide it by 162°, gives 138°, whereas if you deduct 50° from 293°, it leaves the increase of heat by the fire 243°. Steam of 60 lbs. to the inch multiplied by 471, being the inches of steam made by 1 inch of water divided by 243°, the degrees of heat raised by the coal, gives a product of 116; therefore, by Mr. Watt's view it appears that low steam would do one-fifth more duty than high steam, and yet Binner Downs engine in actual work performs about four times the duty given by Mr. Watt's theory and practice, with only one-sixth part of the amount of heat carried off by the condensing water, proving that high steam has much less heat, in proportion to its effective force; and this is further proved by the small quantity of condensing water required to extract its heat.

"Yesterday I proved this 70-inch cylinder while working with the fire-flues round it, which flues only consumed 5 bushels of coal in twenty-four hours. The engine worked eight strokes a minute, 10-feet stroke, 11 lbs. to the inch effective force on the piston; steam in the boiler 45 lbs. above the atmosphere, consuming 12 bushels of coal in four hours, using 13 gallons of condensing water at each stroke, which was heated from 70° to 104°; but when the fires round the cylinder were not kept up, though still having the casing of hot brickwork around it, and performing the same work, burnt 17 bushels of coal in the same time of four hours, and required 15½ gallons of condensing water, which was heated from 70° to 112°. You will find that the increased consumption of coal, by removing the fire from around the cylinder, was nearly in the same proportion as the increase and temperature of the condensing water, showing the experiment to be nearly correct.

"From the general reports of the working of the engines it appears that when the surface sides of the castings are heated, either by hot air or high steam, the duty increases nearly fifty per cent. from this circumstance alone.

"A further proof of the more easy condensation of high steam was in the Binner Downs 42-inch cylinder engine, 9-feet stroke, six strokes per minute, 11 lbs. effective power on each inch, burning 1⅓ bushel of coal an hour. In this engine the proportion of saving by the heating flues was the same as in the large engine. I tried to condense the steam by the cold sides of the condenser, without using injection-water. The water in the condenser cistern was at 50°. After working for twenty-five minutes the small quantity of hot water discharged at the top of the air-pump reached 130° of heat, but then would rise no higher, the cold sides of the condenser being equal to the condensation of all the steam. The eduction-pipe and air-pump, with its bottom and top, gave 60 feet of surface sides of thick cast iron, and about 20 feet more of surface sides of a thin copper condenser; altogether, 80 feet of surface cold sides, surrounded by cold water. About half a pound on the inch was lost in the vacuum, the discharged water being 130° of heat instead of 100°. The vacuum was made imperfect by about $1\frac{1}{2}$ lb. to the inch.

" It is my opinion that high steam will expand and contract with a much less degree of heat or cold in proportion to its effect, than what steam of atmosphere strong will do. I intend to try steam of five or six atmospheres strong, and partially condense it down to nearly one atmosphere strong, and then by an air-pump of more content than is usual to return the steam, air, and water, from the top of the air-pump, all back into the boiler again, above the water-level in the boiler, and by a great number of small tubes, with greatly heated surface sides, to re-heat the returned steam; though by this plan I shall lose the power of the vacuum, and also the power required on the air-bucket to force the steam and water back again into the boiler, yet by returning so much heat I shall over-balance the loss of power, besides having a continued supply of water, which in portable engines, either on the road or on the sea, will be of great value.

" I shall esteem it a very great favour if you will be so good as to turn over in your mind the probable theory of those statements, and give me your opinion. If Mr. Watt's reports of his experiments are correct, how is it possible that the high-pressure engine that I built at the Herland thirteen years ago, which discharged the steam in open air, did more than twenty-eight millions? If you wish, I will send a copy of the certificate of the duty done by this engine, which states very minutely every circumstance. Now that cylinder, with every part of the engine, was exposed to the cold; had it been heated around those surfaces, as on the present plan, it would have done above forty millions.

" Suppose the Binner Downs 70-inch cylinder engine, 10-feet stroke, working with full steam to the bottom of the stroke, when, by the experiment, the heated flues were again laid on would have worked one-third expansive, by the heat of 5 bushels of coal around the cylinder. Now one-third of the power would make a 3 feet 4 inch stroke, 11 lbs. to the inch effective power, eight strokes a minute, during twenty-four hours, by the consumption of 5 bushels of coal applied on the surface sides of the cylinder, performing a duty of 324 millions with a bushel of coal. Now suppose the cylinder without the heating flues had the steam cut off at two-thirds of the stroke, and that it is possible in a moment to heat the cylinder by the flues; in that case the steam would, by its expansion from the hot sides, fill the last third of the cylinder to the bottom of the stroke; then if that steam could be suddenly cooled, so as to contract it one-third, the piston would ascend one-third its stroke in the cylinder; and it appears in theory by this plan, that a cylinder once filled two-thirds full of steam, by receiving the heat on its surface sides from 5 bushels of coal, and again suddenly cooling down, would continue to work for ever, without removing the steam from the cylinder, and would perform a duty of 324 millions. This never can be accomplished in practice in this way, but the effect may be obtained by partially condensing in a suitable condenser, and again heating by hot sides.

" This mystery ought to be laid open by experiment, for what I have stated are plain facts from actual proofs, and I have no doubt that time will show that the theory of Mr. Watt is incorrect. Though there were 300 feet of cold sides, yet 200 feet were not condensing steam, because on the return of the piston, what was condensed below, and while the engine was resting, did not make against it more than what was condensed above the piston on its descent; therefore you may count on 150 feet of cold external sides constantly condensing, that made this third-part difference against the expansion of the steam.

" I remain, Sir,

" Your very humble servant,

" RICHD. TREVITHICK."

The writer's note-book used during those experiments is in his possession, as well as Trevithick's note-book giving particulars of experiments at several mines, from which the following extracts are taken:—

"CORNWALL, *August*, 1828.—Wheal Towan 80-inch cylinder, 10-feet stroke, 6·9 strokes per minute, loaded to 9·5 lbs. on the inch of the piston, with three of Trevithick's boilers, each 37 feet long, 6 feet 2 inches diameter, with fire-tube 3 feet 9 inches diameter, fire-place 6 feet long, evaporated 13 square feet of water with 1 bushel of coal,[1] duty 87 millions. The heat in the stack was just the same as the heat of the steam in the boiler. Another engine of the same size on the same mine, with similar boilers, but working only 4·06 strokes per minute, loaded to 4·55 lbs. on each inch of the piston, did 50·8 millions.

"Wheal Vor 53-inch cylinder, 9-feet stroke, 6·59 strokes per minute, loaded to 19·58 lbs. on each square inch of the piston, did 36·6 millions.

"Wheal Damsel 41-inch cylinder, 7 feet 6 inch stroke, 5·52 strokes per minute, loaded to 21·5 lbs. on the inch of the piston, did 33 millions.

"It would appear, therefore, that about 10 lbs. to the inch on the piston allows of the best duty, and that a 10-feet stroke exceeds in duty a 7 feet 6 inch stroke.

"The Wheal Towan engine, doing 87 millions, had 1248 feet of tube fire-surface, and a similar amount of external boiler surface in the flues. 2½ bushels of coal were consumed each hour, giving about 1000 feet of fire-sides for each bushel of coal consumed per hour, and 50 feet of fire-bars. Those boilers were intended to supply steam for working the engine at ten strokes a minute; a bushel of coal an hour would in that case have had 600 feet of boiler fire-surface.

"Binner Downs 70-inch cylinder, 10-feet stroke, did 41 millions. A fire was then put around the cylinder and steam-pipes, which burnt 5 bushels of coal in twenty-four hours, by which the duty was increased to 63 millions. The surface sides of the

---

[1] 84 lbs.

cylinder, cylinder-top, and steam-pipes heated by flues was 300 feet, and caused a saving of 41 bushels of coal in twenty-four hours. Another engine in the same mine was tried, having a 42-inch cylinder; when the fire was around the cylinder, she worked 100 strokes without injection-water; the expansion-valve was closed at half-stroke, the steam in the boiler 56 lbs. on the inch above the atmosphere."

It is not easy to deal with the important reasonings flowing from those facts, and influencing the form and economy of the steam-engine, nor to show if Trevithick was right in discrediting the laws laid down by Watt. Newcomen's engine had the interior, as well as the exterior of the steam-cylinder exposed to the cooling atmosphere. Watt, by putting a cover on the cylinder, reduced the loss of the heat from the interior, and by his steam-case hoped to reduce the loss from the exterior, though by it he increased the amount of surface exposed to the cold. In Trevithick's early engines the boiler alone exposed heat-losing surface, and this was further reduced by its own comparatively small size, the engine and boiler complete not exposing one-quarter of the surface of a Watt low-pressure engine of equal power. One object of the Binner Downs experiment was to further curtail this loss of power by increasing the heat of the steam while in operation in the cylinder, since called superheating steam.

This principle of giving increased heat to steam, after it had left its state as water, was made practical by Trevithick's boiler at Wheal Prosper in 1810, where the flues having first been carried around the water portion of the boiler, then passed over the steam portion;[1] and again in the upright boiler of 1815, having the upper end of the fire-tube surrounded by steam above the water line.[2] Those early beginnings of super-

---

[1] See vol. ii., p. 71.   [2] See vol. i., p. 354.

heating steam and surface condensation culminated in the Binner Downs experiments of 1828, one immediate practical result of which was the tubular surface condenser, enabling steamboat boilers to avoid, in a great measure, the use of salt water, facilitating in a marked degree the application of marine boilers and engines with steam of an increased pressure.

The Binner Downs engine, with a cylinder of 70 inches in diameter, and a stroke of 10 feet when working with steam in the boilers of 45 lbs. to the square inch above the atmosphere, and using the heating flues around the cylinder, required 13 gallons of injection-water at each stroke, and consumed at the rate of 3 bushels of coal an hour, to produce a duty equal to eighteen millions; by removing the cylinder superheating flues, the quantity of injection-water for the same amount of work increased to $15\frac{1}{2}$ gallons, and the coal to $4\frac{1}{4}$ bushels. Watt's rule for his low-pressure steam vacuum engine doing a duty of eighteen millions, gave 57 gallons of injection-water, and $11\frac{1}{4}$ bushels of coal.

On the question of coal, this statement agrees very nearly with Trevithick's letters of sixteen years before, when he used the high-pressure boilers in the Dolcoath pumping engine,[1] promising that his high-pressure expansive engine would do the work with one-third of the coal required in the low-pressure vacuum engine.

The high-pressure steam required a less amount of injection-water to condense it than the low-pressure steam, in proportion to the work done, showing the Watt rule and the Watt experience to be inapplicable to high-pressure engines; for instead of 57 gallons of injection-water the Binner Downs engine with steam of 45 lbs. to the inch required but $15\frac{1}{2}$ gallons of injection-

---

[1] See vol. ii., p. 171.

water, and this amount was further reduced to 13 gallons by superheating the steam ; this roughly agrees with the coal consumed, or in other words, with the amount of heat to be carried off by injection-water : the Watt rule giving 11¼ bushels as the fair allowance for low-pressure steam vacuum engines, while the high-pressure steam vacuum engine burnt but 4¼ bushels. This was further reduced to 3 bushels by superheating. Those facts led to the idea that if the steam pressure was sufficiently increased, condensation might be carried out without any injection-water, by the transmission of the heat in the steam through the metal sides of the condenser. An experiment was at once made by removing the Watt condenser and injection-water, as he had done seventeen years before,[1] using in their stead a thin copper surface-condenser immersed in cold water, producing, within ½ lb. on the inch, as good a vacuum as when injection-water was used, leading to the conclusion,—

" It is my opinion that high steam will expand and contract with a much less degree of heat or cold, in proportion to its effect, than what steam of atmosphere strong will do. I intend to try steam of five or six atmospheres strong, and partially condense it down to nearly one atmosphere strong, and then by an air-pump of more content than is usual to return the steam, air, and water back into the boiler again, and by a great number of small tubes, with greatly heated surface sides, to reheat the returned steam."

This, in practical words, is the surface condenser by which the used steam is returned to the boiler in the form of water. The more general use of high-pressure steam of 70 or 90 lbs. to the inch, increasing its expansive force on one side of the piston by superheating it

---

[1] See Query 3rd, vol. ii., p. 19.

on its passage through numbers of small tubes, and decreasing its expansive force on the other side of the piston by cooling it in passage through similar tubes exposed to cold, is partly effected in steamboats, but has not yet been attempted in engines on the road.

After a month's further consideration he wrote :—

"Wheal Towan engine is working with three boilers, all of the same size, and the strong steam from the boilers going to the cylinder-case; the boilers are so low as to admit the condensed water to run back from the case again into the boiler: they find that this water is sufficient to feed one of these boilers without any other feed-water, therefore one-third of the steam generated must be condensed by the cold sides of the cylinder-case, and this agrees with the experiments I sent to you from Binner Downs. Wheal Towan engine has an 80-inch cylinder, and requires 72 bushels of coal in twenty-four hours, therefore, the cylinder-case must, in condensing high-pressure steam, use 24 bushels of coal in twenty-four hours. Boulton and Watt's case for a 63-inch cylinder working with low-pressure steam, condensed only 4½ bushels of coal in equal time, the proportions of surface being as 190 to 240 in Wheal Towan. Nearly five times the quantity was condensed of high steam than of low steam, proving that there is a theory yet unaccounted for." [1]

These apparent facts are, in the case of steamboats, more culpably overlooked now than when he wrote forty-two years ago; engines have been examined and reported on by eminent scientific men, but it was left for Trevithick to point out that cold on the surface of the steam-case of a Watt low-pressure steam vacuum engine condensed about one-fifteenth of the steam given from the boilers, and that the loss from exposure to cold was nearly five times more from high-pressure steam than from low-pressure. Within a few more months he

---

[1] See Trevithick's letter, January 24th, 1829, vol. ii., p. 368.

determined on constructing an engine for the purpose of more accurately testing those views.

"Mr. Gilbert,   "Hayle Foundry, *July 27th*, 1829.

"Sir,—Below you have a sketch of the engine that I am making here for the express purpose of experimenting on the working the same steam and water over and over again, heating

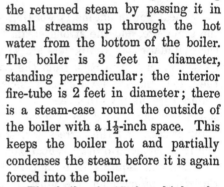

the returned steam by passing it in small streams up through the hot water from the bottom of the boiler. The boiler is 3 feet in diameter, standing perpendicular; the interior fire-tube is 2 feet in diameter; there is a steam-case round the outside of the boiler with a 1½-inch space. This keeps the boiler hot and partially condenses the steam before it is again forced into the boiler.

"The boiler is 15 feet high; the cylinder 14 inches diameter, with a 6-feet stroke, single power. The pump for forcing the steam and water back again is 10 inches in diameter, with a 2 feet 9 inch stroke, about one-quarter part of the content of the steam-cylinder. The bottom of the boiler will have a great number of small holes, about $\frac{1}{16}$th of an inch in diameter, through which the steam delivered into the boiler will pass up through the hot water, by which I should think it will heat those small streams of steam again to their usual temperature.

"The pump for lifting water to prove the duty of the engine is 30 inches in diameter, with a 6-feet stroke, but this may be lengthened to a 12-feet lift, as the trial or load in the experiments may require, giving from 12 to 24 lbs. to the inch in the piston.

This machine will be ready before your return to Cornwall, and I intend to prove it effectually before I go to Holland.

"The Holland engine lifted on the trial, when they came down to see it, 7200 gallons of water a minute 10 feet high with 1 bushel of coal an hour; exceedingly good duty for a small engine of 24-inch cylinder, being 34,560,000 of duty.

"On the 17th August the trial comes on between the two companies about the quays. They are as desperate as possible on both sides, and castings and every other article are thrown down to 30 per cent. below cost price; iron pumps for 6s. 6d. per cwt., and coal sold to the mines for 37s. 3d. per wey, when 48s. per wey on board ship was paid for it. Several thousands lost per year by each party. This never can last long. If you can think of any improvement I shall be very glad to bear in time, before it may be too late to adopt it. At all events, if it is not too much trouble to write, I shall be very glad to hear from you. What effect do you think the water will have in heating the steam on its passage to the top of the water from the false bottom of the boiler?

"I have a cistern of cold water, with a proper condenser in it, connected between the bottom of the boiler-case and the force-pump to the bottom of the boiler, therefore I can partially condense by cold water sides, or by cold air sides just as I please, by rising or sinking the water in the cistern.

"The boiler is made very strong to try different temperatures, and an additional length to the water-pump makes all very suitable for a great number of experiments, and if there is any good in the thing I will bring it out.

"I shall have indicators at different places to prove what advantages can be gained. I hope to have the pleasure of your company during those experiments, which I think will throw more light on this subject than ever has yet been done. Some trials since I last wrote to you make me very confident that much good will arise from these experiments, but to what extent is uncertain.

"I remain, Sir,

"Your most obedient servant,

"RICHARD TREVITHICK."

Trevithick did not use letters to illustrate his sketch, knowing that Davies Gilbert would comprehend it; but the reader of to-day may not find it so easy, therefore the writer has added them with a slight detail description, he having been Trevithick's daily companion when those drawings and experiments were made. *a*, top of boiler; *b*, water line; *c*, centre of wheel; *d*, cast-iron wheel and chain; *e*, chimney, 13 in. in diameter; *f*, fire-tube, 2 ft. diameter; *g*, outer boiler-case, 3 ft. diameter, 15 ft. long; *h*, water space of 6 in.; *i*, boiler steam-case, 3 ft. 4 in. diameter; *j*, small holes through which steam and water are forced into the boiler; *k*, force-pump, 10 in diameter, 2 ft. 9 in. stroke; *l*, steam-cylinder, 14 in. diameter, 6-ft. stroke; *m*, piston-rod; *n*, fire-door; *o*, fire-bars; *p*, pump for testing the power of the engine.

There is a natural tendency in men of genius to unwittingly return, under new forms, to old ideas. The ideas are similar, though in combination with new forms and new acquirements; even the outline of this 1828 boiler, with the exception of its outer steam-casing, is very like that in a letter to Davies Gilbert fourteen years before,[1] of which Trevithick had kept no copy. When in the foregoing letter he wrote, "There is a steam-case round the outside with a 1½-inch space; this keeps the boiler hot and partially condenses the steam before it is again forced into the boiler," he had forgotten that twenty-seven years before, when constructing his first high-pressure steam-engines, he thus specified his invention:—"The steam which escapes in this engine is made to circulate in the case round the boiler, where it prevents the external atmosphere from affecting the temperature of the in-

[1] See Trevithick's letter, 7th May, 1815, vol. i., p. 364.

cluded water, and affords by its partial condensation
a supply for the boiler itself." [1]

Not one of his numerous patent specifications has
been found among his papers, neither do his letters
refer to them ; probably he never read them after the
first necessary examinations.

" Mr. Gilbert,   " Hayle Foundry, *November 5th*, 1829.

"Sir,—The engine has been worked. The result is ten
strokes per minute, 6-feet stroke, with half a bushel of coal per
hour, lifting six thousand pounds weight. This was done with
water in the cistern round the condenser, which water came up
to 180 degrees of heat, and remained so. The water sides of
the condenser covered with this hot water was 50 surface feet.
I tried it to work with the cold air sides, but I found that the
cold air sides of 120 feet would only work it four strokes per
minute. I should have worked the steam much higher than
50 lbs. to the inch, but being an old boiler I thought it a risk.
I am now placing an old boiler of 350 feet of cold sides more to
the condenser, to give a fair trial to condensing with cold sides
alone. The steam below the piston was about 6 or 7 lbs. to
the inch above the atmosphere. The force-pump to the boiler
was about one-fifth part of the content of the cylinder, and the
valve close to the boiler lifted when the force-piston was down
about two-thirds of its stroke, at which time the returned steam
entered the boiler again. I have no doubt of doing near ten
times the duty that is now done on board ships, without using
salt water in the boiler, as at present. Our boiler has been
working three days and the water has not sunk 1 inch per day.
I am quite satisfied the trial will be a great success.

" Mr. Praed and Sir John St. Aubyn are anxious to get a high
bank carried out from Chapel Angel to 15 feet below low-water
mark on the bar, to make Hayle a floating harbour.

"I have proposed to make a sand-lifting engine. When I
built that engine for deepening Woolwich Harbour, we lifted
300 tons per hour through 36 feet of water, and 20 feet above

[1] See patent specification of 1802, vol. i., p. 128.

water, 56 feet above the bottom. This was done with two bushels of coal per hour, therefore it will not cost above one penny per square fathom to lift the sand over this embankment. It is intended to get down Mr. Telford to give his opinion on it. Your remarks on it would be of service.

"I remain, Sir,

"Your humble servant,

"RICHARD TREVITHICK."

The writer having worked at these experiments, knows that their object was to employ high-pressure steam in the boiler, using it very expansively in the cylinder, and by cold surface sides reducing its bulk either to low-pressure steam or boiling water, and then force it again into the boiler.

" MR. GILBERT,   " HAYLE FOUNDRY, *November 14th,* 1829.

" Sir,—I have both of your letters and sketches, which shall be put in hand. I understand it perfectly well. Since I wrote to you last I have made several satisfactory trials of the engine, and think it unnecessary to make any further experiments. The statement below may be depended on for a future data. The load of the engine was 6280 lbs., being 20 lbs. to the inch for a 20-inch cylinder with a 6-feet stroke, 12 strokes per minute, with three-quarters of a bushel of coals per hour, giving a duty of 361,728,000 for 1 bushel of coal, a duty far beyond anything done in the county by so small an engine. The cold water sides round the condenser was 60 feet, and the water at 112 degrees temperature, not having a sufficient stream of cold water to supply the cistern. Each foot of cold water sides did 7536 lbs. per minute, about three times the work done in the county per foot of hot boiler sides; therefore the condenser need not be more than one-third of the boiler sides. By making the condenser of 4-inch copper tubes and of an inch thick, it would stand in one-twentieth part of the space of the boiler.

" I put a boiler naked to try cold air sides; it was very rusty, and did not condense as fast as I expected. The engine worked exceedingly well, but slow. The duty performed for each foot of cold air sides was 565 lbs. per minute, about one-thirteenth part of the condensing of cold water sides. We never wanted to get the steam above 60 lbs. to the inch. I have no doubt but that copper pipes of $\frac{1}{32}$nd of an inch thick, clean and small, would do considerably more, because the hot water that came out of the boiler from the condensed steam was but 170 degrees, and the external sides the same heat when the steam was 15 lbs. above the atmosphere in the condensing boiler. This boiler was 4 feet 6 inches diameter, and I think that towards the external sides of the boiler there was a colder atmosphere, if I may call it so, than what it was in the middle of this large condensing boiler, because I found by trying a small tin tube, that it would condense 1500 lbs. for each foot of cold air sides.

"However, as it is, it will do exceedingly well for portable purposes.

"The duty, I doubt not, will be, both for water and air sides condensing, at least 50 per cent. above our Cornish engines, which will be above four times what is now done with ships' engines, especially when you take into consideration their getting steam from salt water, and letting out so much water from the boiler to prevent the salt from accumulating in the boiler, which will make 30 per cent. more in its favour.

"If strong boilers to stand 200 lbs. to the inch are made with small tubes, I have no doubt but that the duty would be considerably more, and my engines will not be one-quarter part of the weight, price, or space of others; and when every advantage is taken it will be 1000 per cent. superior in saving of coal to those now at work on board. This engine works well, and returns the steam very regularly every stroke into the boiler.

"I am extremely sorry you were not present to see these experiments. Please make your remarks on these statements, with any further information you may judge useful.

"I shall now make drawings agreeable to my experiments for actual performance on board ships. In hope of hearing from you soon,

"I remain, Sir,

"Your very humble servant,

"RICHD. TREVITHICK."

The large old boilers used as surface condensers, in which the steam was partially condensed by the transmission of heat to the external atmosphere, together with its further condensation in a smaller condenser with cold water around it, so reduced its expansiveness, that a large feed-pump drew the hot water and steam from the small condenser, and forced it back into the boiler without any reduction of quantity; those temporary contrivances, almost immediately resolved themselves into a condenser made of copper tubes surrounded by cold water.

Having proved by six months' experiment on a working scale the practicability of the plan which in reality he had invented twenty years before in the iron steamship,[1] he wrote in June, 1830 :—

"To the Right Honourable the Lords Commissioners
of the Admiralty, &c., &c., &c.

"My Lords,

"About one year since I had the honour of attending your honourable Board with proposed plans for the improvement of steam navigation, and as you expressed a wish to see it accomplished, I immediately made an engine of considerable power for the express purpose of proving by practice what I then advanced in theory. I humbly request your lordships will grant me the loan of a vessel of about 200 or 300 tons burthen, in which I will fix at my own expense and risk an engine of suitable power to propel the same at the speed required: no alterations whatever in the vessel will be necessary. When under sail the propelling apparatus can be removed, and when propelled by steam alone, the apparatus outside the ship will scarcely receive any shock from a heavy sea. This new invention entirely removes the great objection of feeding the boiler with salt water."

This petition was backed by Mr. Gilbert and Mr.

---

[1] Drawing of iron steamship, vol. i., p. 336.

George Rennie. His old friend Mr. Mills took an interest in it, and wrote, " I am going to meet Captain Symonds at Woolwich again to-morrow, and hope to be able to persuade him to use his influence with Sir T. Hardy."

"LAUDERDALE HOUSE, HIGHGATE,

" MR. GILBERT,   "*August 19th*, 1830.

" Sir,—The boiler with the fire-place, cold air tubes outside the boiler but within the steam-case, fire-tubes in the boiler

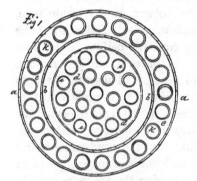

FIG. 1.—PLAN SECTION.

from the top of the fire-place to the top of the boiler, the ash-pit close, except a small door to clear out the ashes.

" The design is for the cold air to pass down from the top of the boiler through the air-tubes within the steam-case surrounding the boilers, becoming heated in its passage by condensing the steam in the case, and then to pass up through

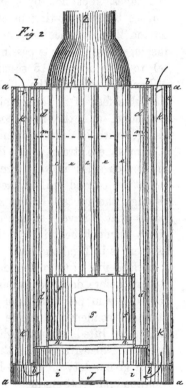

FIG. 2.—ELEVATION SECTION.[1]

the fire-bars in the hot state, nearly as hot as the steam in the case; because this air, heated to nearly 212 degrees by

[1] *a*, steam-case; *b*, boiler-case; *c*, space for condensation of steam; *d*, water and steam space; *e*, fire-tubes; *f*, fire-box; *g*, fire-door; *h*, fire-bars; *i*, ash-box; *j*, ash-box door; *k*, air-tubes in condenser; *l*, chimney; *m*, water level; a smoke-jack fan draught.

condensing the steam in its passage without any of its oxygen being burnt, it will not carry off so much heat from the fire as cold air would, and still have the same oxygen as cold air to consume the coal.

" The cold air will be passing down the steam-case in the air-tubes, and up through the fire and fire-tubes in the boiler. I find by experiments I have made here, by placing a tin tube $2\frac{1}{2}$ inches in diameter, 4 feet long, inside a 4-inch tube of the same length, having boiling water and steam between the tubes, kept hot by a fire round the outer tube, with a smith's bellows blowing in at the bottom of the inside tube, having $2\frac{2}{3}$rds surface feet of condensing sides, measuring the inside, where the air is passing up from the bellows, heats from 60 to 134 degrees 15 square feet of cold air per minute. When you compare the effective heat of 74 degrees given to 15 cubic feet of air every minute from $2\frac{2}{3}$rds surface feet of tin plate, and the heat contained in 15 cubic feet of air charged with 74 degrees of effective heat, compared with steam of atmosphere strong, you will find that the condensing power of surface sides is very great, and for locomotive purposes might be carried still further, by forcing the air more quickly through the tubes. If the statements on air given in some books that I have read are correct, that there is about three times as much heat in 1 gallon of steam of atmosphere strong as there is in 1 gallon of air of 212 degrees of heat, in that case 1 surface foot of tin-plate sides of this pipe, by sending off the hot air before described, would take out the heat of $1\frac{1}{2}$ cubic foot of steam per minute of atmosphere strong, which in the common condensing engine would be equal to a duty of 2700 lbs. lifted 1 foot high per minute; but in the high-pressure expansive engine, the heat of $1\frac{1}{2}$ cubic foot of steam would give a duty of 10,800 lbs., or four times the duty of the Boulton and Watt engine.

" If you calculate on the air being heated to nearly 212 degrees before it enters the fire, together with the heat given to the sides of the boiler, the fuel saved will be above one-half on what has been done by the high-pressure engines in Cornwall, because at present the coal must pay for heating the cold air, therefore a less proportion goes through the sides of the boiler, and is lost through the chimney; whereas if the heat of the steam, by passing into the cold air, on its way through the condenser tubes, is carried into the fire-place, one-half of the coal must be saved; and you will find by calculation that the quantity of air required to burn the coal, and also to condense the steam, goes exactly in proper proportion for each other, and for locomotive engines with a blast will go hand-in-hand almost to any extent, and the size of an engine, for its power, is a mere nothing.

"A smoke-jack fan in the ash-pit under the fire-bars, worked by the engine, would draw air down the condensing tubes, and force it up through the fire and fire-tubes always with the speed required, as the steam and the condensation would increase in the same ratio.

"As it is possible to blow so much cold air into a fire as to put it out, by first heating the air it would burn all the stronger, and whatever heat is taken out of the condenser into the fire-place from the steam that has been made use of, half this extra heat will go into the boiler again, or in other words, but half the quantity of cold will be put into the fire, being the same in effect as saving fuel. Taking heat from the condenser through the boiler sides is an additional new principle in this engine. I find by blowing through tubes that the condensation of a surface foot of air-tube against a surface foot of boiler fire-tube is greater than the fire that passes through the boiler sides, where the common chimney draught is used, by nearly double; but I expect when both air and fire tubes are forced by a strong current of air it will be nearly equal, and the increase of steam and of condensation can be increased by an increased current of air, so as to cause a surface foot of fire and of air sides to do perhaps five times as much; and of course the machine will be lighter in proportion. I think air sides condensation preferable to water sides, as so small a space does the work, and is always convenient, and its power uniformly increasing with its speed, by the increased quantity of air, without the weight of water vessels. This kind of engine can be made to suit every place and purpose, and I think such an engine of the weight of a Boulton and Watt engine will perform twenty times the duty.

"Air sides condensation will be advantageous on board ship, because there are holes for the passage of water through the bottom and sides of the ship.

"I am anxious to have your opinion on this plan of returning the hot air from the condenser to the fire-place, and what you think the effect will be.

"The Comptroller of the Navy has not yet returned from Plymouth, therefore no answer has been given to me.

"You will see by the sketch how very small and compact an engine is now brought without complication or difficulty; each surface foot of boiler and condenser is equal to one-third of a horse-power, weighing 20 lbs., or 60 lbs. weight for each horse-power. The consumption of fuel is so small when working a differential engine, that I expect it will not exceed 1 lb. of coal per hour for each horse-power.

"The cost of erection and required room are so small from its simplicity that it will be generally used. As I am very anxious that every possible improvement should be considered prior to making a specification for a patent, I must beg that you will have the goodness to consider and calculate on the data I have given you. I am sorry to trouble you, but I am satisfied this will be to you rather a pleasing amusement than a trouble. The warming machines will take a very extensive run, and I believe will pay exceedingly well.

"I am almost in the mind to take a ride down to see you in a few days, but am now detained here about the American mining concerns.

"I remain, Sir,

"Your very humble servant,

"RD. TREVITHICK."

The letters and foot-note are the only changes made by the writer in Trevithick's original sketch so descriptive of a wonderful application of varied and improved principles of long-known difficulty and importance; the beautifully compact tubular boiler for giving high-pressure superheated steam, surface condensation, absence of feed and condensing water, and return of the heat, in other engines wasted in condensation, to the fire-place; though there is little or no mention of the mechanical or moving parts of the steam-engine, yet its vital principles are grasped with the hand of a master. The sketch in the letter hastily made forty years ago is more ingenious than any portable engine since constructed, though there may be no sufficient proof of its practical success. The propeller to be worked by this novel engine was of course his long-idle screw.

Steam Engines, 21st February, 1831.

"Now KNOW YE, that in compliance with the said proviso, I, the said Richard Trevithick, do declare that the essential points in my improved steam-engine, for which I claim to be the first and true inventor, are :—

"Firstly, the placing of the boiler within the condenser, in order to obtain the additional security of the strength of the condenser to prevent mischief in case the boiler should burst, and also by the same arrangement to conveniently make the condenser, with a very extensive surface, enabling me to condense the steam without injecting water into it.

"Secondly, the enclosing of the condenser in an air or water vessel, by which the intention of safety from explosion is further provided for, and my engine really rendered what I denominate it, a high-pressure safety engine.

"Thirdly, the condensing of the steam in the condenser by means of a current of cold air or cold water forced against the outsides of the condenser.

"Fourthly, the returning of the condensed steam from the condenser back again into the boiler, to the end that sediment and concretion in the boiler may be prevented ; and,

"Fifthly, the blowing of the fire with the air after it has been heated by condensing the steam.

"In forming my improved steam-engine I employ several or

2 A 2

all of these points according to convenience, in combination
with the other necessary parts of steam-engines in common use.

"These, my essential points, will admit of various modifications as to form and proportions such as must be and are quite
familiar to every competent steam-engine manufacturer, and
therefore it will be sufficient for the perfect description of my
improved steam-engine that I explain some of the modes of
forming and combining the essential points of my invention
with the other parts of steam-engines in common use. In my
most favourite form of engine in which I condense by a current
of cold air, the fire-place and flue, the boiler, the condenser, and
the air-vessel, are made of six concentric tubes, standing in an
upright position. The inner or first tube forms the fire-place
and flue, and at the same time the inner side of the boiler.
This tube is conical, having its small end upwards. The next
or second tube is cylindrical, about 6 inches larger in diameter
than the lower end of the first tube, and forms the outside of
the boiler, leaving a space all round of about 3 inches at the
bottom, and so much more at the top, as the flue is taper for
holding water and steam between the two tubes. The third
tube is about 2 inches larger in diameter than the second, in
order to allow a space of about an inch for powdered charcoal or
some other slow conductor of heat. This tube also constitutes
the inner side of the air-vessel. The fourth tube is about
2 inches larger than the third, and forms the inner side of the
condenser. The fifth tube, about 2 inches larger than the fourth,
forms the outside of the condenser; and the sixth tube, about
2 inches larger than the fifth, forms the outside of the air-vessel,
and at the same time the outside of the whole of the generating
and condensing apparatus, consisting of fire-place, flue, boiler,
condenser, and air-vessel. These tubes are made of wrought-
iron plates riveted together, and are all cylindrical, except the
first, which is conical, the bottom or fire end being the largest.
The first or inner tube is closed at bottom, but has an opening
on one side near the bottom, through which the fire-bars are
introduced, and the ashes and clinkers taken away. To this
opening a neck-piece about 3 inches long is riveted, having a
flange to fit against the inside of the second tube, when the two

tubes are concentric, through the side of which second tube is
an opening corresponding with that in the first tube, and the
flanch is screwed to the second tube so as to make one opening
through the sides of the two tubes. The second tube extends
downwards about 5 inches below the first tube, and has a flanch
turning inwards, to which a second round plate of iron is
screwed, forming the bottom of the boiler. The first tube has
an external flanch at the top, and the second tube an internal
flanch, both of the same height, and screwed to a cast-iron
circular plate or cap-piece, which extends wide enough around
the boiler to form also the cover for the air-vessel. This plate
has a hole in the middle as large as the flue. The sides of the
condenser and air-vessel are formed of four concentric tubes,
each about 2 inches larger than the one within it. The inner
and outer of these tubes constitute the sides of the air-vessel,
and are each furnished with an external flanch at the top by
which they are screwed to the cap-piece. The two intermediate
tubes constituting the sides of the condenser are riveted
together at the top, leaving a space of about an inch between
their upper ends and the cap-piece, so as to allow of a free
communication over them between the outer and inner parts of
the air-vessel. The inner tube of the air-vessel extends downwards about an inch below the boiler, and is closed by a flat
plate screwed on to a flanch projecting inwards from the tube;
the two tubes of the condenser descend about 3 inches lower
than the boiler. The inner tube has an internal flanch, to
which a flat circular plate is screwed to close up the tube.
The outer tube of the condenser is of the same length with the
inner, and is provided with an external flanch about 3 inches
broad. The outer tube of the air-vessel has an external flanch
2 inches broad, and is just long enough to come down upon the
broad flanch of the condenser last described, and these two
flanches are together bolted upon a bottom piece of cast iron,
which is a dish of 4 inches deep, and equal in diameter with
the diameter of the outer tube, and having a flanch the same
breadth as the flanch of the outer tube, and the bottom piece
is secured to the air-vessel and outer tube of the condenser by
bolts going through all the three flanches. An opening is

made through the sides of all the four tubes of the condenser
and air-vessel opposite to and as wide as the fire-place opening
through the sides of the boiler. The upper part of both
openings to be of the same height, but the outer opening is
made as low as the bottom of the boiler, in order to allow room
for a pipe to enter that part of the boiler for forcing the water
into it, and also another pipe and cock for drawing off the water
or sediment, in case foul water be used by accident or carelessness. These two openings through the condenser and air-vessel,
and through the boiler, constitute one fire doorway through all
the six tubes for access to the fire-place; a ring is placed
between the two tubes of the condenser around the fire doorway, so as to cut off all communication of the steam in the
condenser with the air in the doorway; another similar ring is
placed between the condenser and the outer tube to prevent
the escape of air into the fire doorway, and a half ring is placed
in the lower part of the fire doorway between the condenser
and the inner tube of the air-vessel, to prevent ashes from
falling into the air-vessel, and yet allow a free passage for the
air from the inner part of the air-vessel into the upper part of
the fire doorway. These two rings and the half ring are
secured in their places by rivets passing through all of them
and through the tubes, and uniting all firmly together, the
interstices being filled with iron cement. A ring is also placed
between the boiler and the air-vessel around the fire doorway,
against the outside of which ring the charcoal powder is tightly
rammed, and will hold the ring in its place without the necessity of either rivets or screws. That part of the fire doorway
which is above the fire-bars is supplied with an inner door, to
shut the fire-place even with the outside of the boiler, and
exclude all access of air to the fire, except through the grating.
The whole of the fire doorway is enclosed by an outer door even
with the outside of the air-vessel, to exclude all air, except that
which comes through the air-vessel; a pipe is fixed in the
bottom or dish-piece leading to a forcing pump to draw the
water out of the condenser and force it into the bottom of the
boiler through the pipe before de cribed. A blowing cylinder
of about ten times the content of the main cylinder is screwed

against the outside of the air-vessel, and opposite to the two outlet valves of the blowing cylinder two apertures are made in the air-vessel, through which the air is forced in. The main cylinder of the engine, of the usual dimensions according to power wanted, is also screwed against the outside of the air-vessel high enough above the blowing cylinder to allow room for the main-crank shaft to work between them. The forcing pump before mentioned is also screwed to the outside of the air-vessel, and thus my improved steam-engine becomes more compact and convenient than any preceding steam-engine. For the purpose of supplying the boiler with distilled water, in case there should be a deficiency in it, a small vessel made of two upright tubes, one within the other, is placed on the cap-piece. The inner tube is of the same diameter as the flue, and forms a continuation of it. The outer tube is about 6 inches larger than the inner, and the space at the top and bottom between the two tubes is closed by two ring-shaped pieces. This vessel may be about 18 inches high; a cock is fixed in the top of this vessel, to which a bent pipe is fastened, leading to and united with a pipe which arises from the top of the con-denser and passes through a hole in the cap-piece, and thus a communication between the supplying vessel and the condenser may be opened or shut at pleasure; another pipe, also furnished with a stop-cock, arises from the vessel, and communicates with a water-cistern to receive its supply of water when required; a third pipe, having a cock in it, opens into the vessel near the bottom to let out the sediment; a small cock to let the air out is also fixed in the top of the vessel, which cock may also be used for letting air out of the condenser. In order to supply the boiler with water by means of this vessel, the stop-cock leading to the condenser is shut, and that leading to the cistern is opened, and at the same time the air-cock is opened to allow the air to escape that the water may fill the vessel. When the vessel is nearly full of water, the air-cock and the cock from the cistern are shut, and that in the pipe leading to the condenser is opened. The water being then heated by the flue is con-verted into steam, which, passing into the condenser, is there reduced to water again, leaving the sediment or salt in the supplying vessel, which sediment or salt may be occasionally blown out through the bottom pipe by filling the vessel with water, shutting the water, steam, and air cocks, and opening the cock of the outlet pipe at a time when the steam in the vessel is strong. But the supply of water from the condenser being always equal to that converted into steam and used in the engine, there is no tendency to a variation in the height of the water in the boiler, except there be leakage or waste of steam in some part of the engine. An upright glass tube, having an iron tube of communication with the lower part of the boiler and another iron tube of communication to the upper part of the boiler, is conveniently placed against the outside of the air-vessel to indicate at all times the height of the water in the boiler; as is usual in steam-boilers, a valve is placed on the top of the air-vessel to allow of the escape of a portion of the air in case that the quality of the fuel should not require so much air for perfect combustion as the steam requires for good condensation. The degree of the condensation of the steam may be increased at pleasure, by increasing the velocity of the air passing into and through the air-vessel. The other parts of my improved steam-engine, such as the steam-pipes, the throttle-valve, the safety-valve, the vacuum-valve, the working valves, crank, connecting rods, cross-heads, pistons, piston-rods, and various other minor parts common to engines in general use may be made in the usual forms, and placed in the most convenient situations; they cannot, therefore, need any description. When it is intended to use water for condensing instead of air, my improved steam-engine must be made as before directed, except that the communication between the air-vessel and the fire-place must be closed, which may be done by a perfect ring of iron surrounding the opening leading to the fire-place, instead of the half ring before described, and a forcing pump must be employed to draw water from a reservoir, and force it into the vessel which I have hereinbefore denominated the air-vessel, but which in this mode of working would more properly bear the name of water-vessel. In this case a blowing cylinder, the dimensions of which must be calculated according to the quality of the fuel to be used, may be worked to blow the fire through a pipe leading into the ash-pit. This, however, will not be necessary where there is a chimney high enough to create a strong draught. In respect to proportions, my improved steam-engine admits of considerable latitude, and it will be sufficient direction to any practical engineer to say that for engines working with steam of 120 lbs. to the inch, used expansively till it be nearly reduced to atmospheric strength and then condensed, a 10-horse engine may have a fire-place of 20 inches diameter, the flue at the top 10 inches diameter, and a boiler of 20 feet high; a 60-horse engine, a fire-place of 36 inches diameter, a flue of 16 inches diameter, and a boiler of 20 feet high. In boat-engines, and in other cases where height cannot be allowed, the diameter must be increased. The thickness of the two tubes constituting the boiler sides of a 10-horse engine may be ⅛th of an inch, that of a 60-horse a quarter of an inch, and so in proportion for engines of other power. The tubes constituting the condenser and inner tube of the air-vessel may in all cases be ⅛th of an inch thick. The outer tube may be ⅜ths of an inch thick, to afford stability to the working cylinder, the blowing cylinder, and the forcing pump fastened to this tube, and as an ultimate perfect barrier against explosion. The respective distances of the other tubes constituting the outside of the boiler, the condenser, and air-vessel, will be the same as hereinbefore given, and therefore their diameters will depend upon the diameter of the fire-place. The cap-piece in small engines may be half an inch thick, and in large engines an inch. The bottom of the ash-pit and bottom of the boiler must have about half an inch of thickness for every foot diameter, or they may be cast with ribs to afford equivalent strength. The fuel is supplied through a door in the flue, at the top of the boiler, consisting of coke or coals the least liable to swell with heat. The flue may be filled to about one-third of the height of the boiler, and the water fill about three-fourths of the boiler, leaving one-fourth for steam.

Having clearly explained my improved steam-engine so that any person competent to make a steam-engine can from this description understand my invention and carry the same into effect in as beneficial a manner as myself, I proceed to observe

that the extreme safety of my improved steam-engine will be
seen, from considering that in case the boiler should explode
inwards into the flue, the power of the steam would be first
reduced by filling the flue and fire-place, and could not escape
through the chimney and fire doorway faster than it would
diffuse itself and be condensed by mixing with the surrounding
air, and thus lose all its force. But should the outside of the
boiler burst, part of the force of the steam would be spent in
filling up the interstices between the particles of the charcoal,
and would then probably be too weak to effect a breach through
the inner tube of the air-vessel; and should such a second
breach be effected, the space within the air-vessel would allow
the steam to expand and partly condense, and a portion to
escape into and through the fire doorway, where it would divide
itself, and proceed harmlessly up the flue, and out at the doorway; so that the outer case being a reserve of strength, would
to a certainty withstand the force remaining in the steam after
the before-mentioned successive reductions of power."

The patent of February, 1831, perfects the sketch in
his letter of July 27th, 1829, which in its turn made
more perfect the plans put into practice in 1815, just
before leaving England for America.[1] The prejudice
against the use of his high-pressure steam-engine he tried
to meet by calling it "a high-pressure safety engine."
The boiler was of six wrought-iron upright tubes, one
within the other. The inner one was the fire-tube, surrounded by a tube of larger diameter, forming the water
and steam space. This was again surrounded by another
tube, 2 inches larger in diameter, the space being filled
with charcoal or other non-conductor of heat; another
tube, 2 inches more in diameter, formed the inner circle
of the condenser, having an inch space for the passage of
cold air from the blowing cylinder, carrying the heat
from the condensing steam back to the fire-place. Still

---

[1] See Trevithick's letters, July 8th, 1815, vol. ii., p. 80, and 7th, vol. i.,
p. 364; and 16th May, vol. i., p. 370; and patent of 1815, vol. i., p. 375.

another tube, 2 inches more in diameter, giving a space into which the used steam from the cylinder passed to be condensed. Then came the outside tube, 2 inches more in diameter, forming a second space for the passage of air, taking heat from the condenser into the fire. The steam-boiler had its heat retained by a coating of charcoal; next to it came a current of cold air an inch thick, carrying back to the fire any heat that had passed through the charcoal coat, and also the heat from the inner surface of the condenser. Then came the inch-thick circle of steam, on its exit from the cylinder, to be condensed; and finally an outside circle of cold air, performing the same functions as the inner circle in condensing the steam and carrying its heat back again to the fire.

The object or principle of this engine was to avoid the loss of heat, and the necessity for either condensing water or feed-water, as described in the letter and drawing of August 19th, 1830, but the detail was changed, mainly to facilitate construction. As in practice it might be impossible to fully attain those objects, preparation was made to get rid of the salt from such water as might be required as feed-water to make good the loss from leakage or other defects in the working of marine steam-engines. The specification states: "For the purpose of supplying the boiler with distilled water, in case there should be a deficiency in it, a small vessel made of two upright tubes, one within the other, is placed on the cap-piece. The inner tube is of the same diameter as the flue, and forms a continuation of it. The water being heated by the flue is converted into steam, which, passing into the condenser, is there reduced to water again, leaving the sediment or salt in the supplying vessel."

352

TUBULAR BOILER, SUPERHEATING STEAM,

Where water condensation was preferred the surface-air condenser could be converted into a surface-water condenser by a current of cold water in place of the air; in which case the air from the blowing cylinder was taken direct in to the fire-place or other means used for giving the necessary draught. Steam of about 135 lbs. to the inch was to be so expansively worked as at the finish of the stroke, on its escape to the condenser, to be no more than atmospheric pressure, or 15 lbs. to the inch—just the strength with which Watt preferred to commence his work in the cylinder.

The most prominent feature in Trevithick's numerous modifications of the steam-engine was the boiler. In the 'Life of Watt,' though his commentators have been numerous and eminent, little or nothing is said about the boiler or the steam pressure. He left that all-important part of the steam-engine just as he found it, resisting the increase of steam pressure, which was the mainspring of Trevithick's engine. The boiler of the high-pressure engines of 1796[1] sheltered the steam-cylinder from cold; and the used steam from the cylinder circulated around the exterior of the boiler, on its way to the blast-pipe, while the condensed portion was returned as feed-water in the patent engine of 1802.[2] In 1811 he proposed to force air into the fire-place, hoping thereby to reduce the amount of heat lost by the chimney.[3] His various forms of tubular boilers, as at the Herland Mine,[4] and at Dolcoath,[5] and the upright multitubular boilers patented in 1815,[6] followed up in 1828. " I shall have a small portable engine

[1] See vol. i., p. 104.
[2] See patent specification, vol. i., p. 128.
[3] See Trevithick's letter, 13th Jan., 1811, vol. ii., p. 6.
[4] See vol. ii., p. 71.
[5] See chap. xx.
[6] See Trevithick's letters, 8th July, 1815, vol. ii., p. 80; and 7th and 16th May, vol. i., pp. 364, 370.

finished here next week, and will try to heat steam independent of water, in small tubes of iron, on its passage
from the boiler to the cylinder, and also try cold sides
for condensing." In 1829 a simple boiler and condenser composed of three tubes was made, the inner or
fire-tube being 2 feet in diameter and 15 feet long, "for
the express purpose of experimenting on the working
the same steam and water over and over again;"[1] and
on the same subject, "By making the condenser of
4-inch copper tubes $\frac{1}{32}$nd of an inch thick, it would
stand in one-twentieth part of the space of the boiler:"[2]
and finally the sketch of the tubular boiler and tubular
condenser of 1830, in its boiler portion similar to the
best portable boilers of the present day, and the patent
specification of 1831. Surely therefore to him belongs
the credit of having invented and perfected the tubular
boiler and surface condenser

Smiles has written :[3]—

"For many years previous to this period (1829), ingenious
mechanics had been engaged in attempting to solve the pro-
blem of the best and most economical boiler for the production
of high-pressure steam. Various improvements had been suggested and made in the Trevithick boiler, as it was called, from
the supposition that Mr. Trevithick was its inventor. But Mr.
Oliver Evans, of Pennsylvania, many years before employed the
same kind of boiler, and as he did not claim the invention, the
probability is that it was in use before his time. The boiler in
question was provided with an internal flue, through which the
heated air and flames passed, after traversing the length of the
under side of the boiler, before entering the chimney.

"This was the form of boiler adopted by Mr. Stephenson in
his Killingworth engine, to which he added the steam-blast with
such effect. We cannot do better than here quote the words

---

[1] See vol. ii., p. 332.   [2] See vol. ii , p. 336.
[3] See ' Life of George Stephenson,' by Smiles, p. 279 ; published 1857.

of Mr. Robert Stephenson on the construction of the 'Rocket' engine:—'After the opening of the Stockton and Darlington, and before that of the Liverpool and Manchester Railway, my father directed his attention to various methods of increasing the evaporative power of the boiler of the locomotive engine. Amongst other attempts, he introduced tubes (as had before been done in other engines)—small tubes containing water, by which the heating surface was materially increased. Two engines with such tubes were constructed for the St. Etienne Railway, in France, which was in progress of construction in the year 1828; but the expedient was not successful; the tubes became furred with deposit, and burned out.

" 'Other engines, with boilers of a variety of construction, were made, all having in view the increase of the heating surface, as it then became obvious to my father that the speed of the engine could not be increased without increasing the evaporative power of the boiler. Increase of surface was in some cases obtained by inserting two tubes, each containing a separate fire, into the boiler; in other cases the same result was obtained by returning the same tube through the boiler; but it was not until he was engaged in making some experiments, during the progress of the Liverpool and Manchester Railway, in conjunction with Mr. Henry Booth, the well-known secretary of the company, that any decided movement in this direction was effected, and that the present multitubular boiler assumed a practicable shape. It was in conjunction with Mr. Booth that my father constructed the 'Rocket' engine.

" 'In this instance, as in every other important step in science or art, various claimants have arisen for the merit of having suggested the multitubular boiler as a means of obtaining the necessary heating surface. Whatever may be the value of their respective claims, the public, useful, and extensive application of the invention must certainly date from the experiments made at Rainhill. M. Seguin, for whom engines had been made by my father some few years previously, states that he patented a similar multitubular boiler in France several years before. A still prior claim is made by Mr. Stevens, of New York, who was all but a rival to Mr. Fulton in the introduction of steamboats on the American rivers. It is stated that as early as 1807 he used the multitubular boiler.

" 'These claimants may all be entitled to great and independent merit; but certain it is, that the perfect establishment of the success of the multitubular boiler is more immediately due to the suggestion of Mr. Henry Booth, and to my father's practical knowledge in carrying it out.'

" We may here briefly state that the boiler of the 'Rocket' was cylindrical, with flat ends, 6 feet in length, and 3 feet 4 inches in diameter. The upper half of the boiler was used as a reservoir for the steam, the lower half being filled with water. Through the lower part twenty-five copper tubes of 3 inches diameter extended, which were open to the fire-box at one end and to the chimney at the other. The fire-box, or furnace, 2 feet wide and 3 feet high, was attached immediately behind the boiler, and was also surrounded with water."

Stephenson knew of Trevithick's patent of 1802,[1] in which a three-tubed boiler is shown; and it was after that time that Oliver Evans and Fulton tried their experiments, and also the numerous engines with single or return double tube, at work in the principal towns of England prior to 1804,[2] and near his residence in childhood and in manhood.[3]

George Stephenson's Killingworth boiler, " to which he added the steam-blast with such effect," was a copy of Trevithick's boiler and blast, working since 1804 in Newcastle-on-Tyne, and was precisely the boiler described by Stephenson; " in other cases the same result was obtained by returning the same tube through the boiler." This is an admission from Stephenson that Trevithick's patent boiler was the best in use up to about 1828.

[1] See vol. i., p. 128.
[2] See Trevithick's letter, Sept. 23rd, 1804, vol. ii., p. 2.
[3] Mr. Armstrong's note, vol. i., p. 184.

A further proof of the indirect public gain from the use of Trevithick's return-tube boiler over a period of thirty years is their having supplied high-pressure expansive steam in the first experiments made with such steam by the Admiralty, at whose request Mr. Rennie and others examined the duty of the Cornish high-pressure expansive engine, and Captain King, R.N., in charge of the Admiralty Department at Falmouth in 1830, gave an order to Harvey and Co. to construct high-pressure steam-boilers for the Government vessel 'Echo'; in 1831 the machinery was put on board the 'Echo' in the Government Dockyard at Plymouth, and included three of Trevithick's return-tube boilers, made of wrought iron, each 5 feet 6 inches in diameter and 24 feet long, with internal return fire-tube 2 feet 2 inches in diameter. The fire-place end of the boiler was 6 feet 9 inches deep by 5 feet 6 inches wide, to give room for the fire-place and ash-pit. The steam pressure was 20 lbs. on the inch above the atmosphere, worked by double-beat valves, 6 inches in diameter, with expansive gear.

This new machinery was fixed under the superintendence of the writer, after which the Government engineers took charge of the vessel, and the writer who had, as the mechanic in charge, worked like a slave, though receiving but 1s. 6d. a day and expenses, was not invited to take any part in the experimental trials, nor ever heard of the result except in the ordinary rumours of Admiralty bungling on board the 'Echo.'

Those boilers were similar to the Trevithick boiler that had served the locomotive in Newcastle and elsewhere from 1801 to 1828, the first steamboat experiments in England, in Scotland, and in America, and the numerous high-pressure engines then at work.

The enlarging the fire-place end of boilers or fire-tubes has led to many forms. Trevithick's model of 1796[1] had an oval tube giving a greater spread of fire-bars; the same is seen in the 1808 steamboat;[2] the Dolcoath boilers of 1811[3] show the oval and also the bottle-neck fire-tube; the Welsh locomotive of 1804[4] had the fire-tube contracted at its bend or return portion; the Tredegar puddling-mill fire-tube of 1801[5] tapered gradually from the fire-bridge to the chimney end; in the London locomotive of 1808[6] the fire-tube took the bottle-neck shape close to the fire-bridge. The accompanying sketch shows the bottle-neck contraction, only on the top and sides of the fire-tube was to give breadth to the fire-bars *d*, and thickness to the fire at bridge *c*, after which the flue portion of the fire-tube was contracted: this boiler was for many years a favourite in Cornwall. The bottle-neck contraction of

BOTTLE-NECK BOILER.

the 'Echo' boiler was similar to the above, except that the enlargement of the fire-place was downwards instead of upwards, and the fire-tube, instead of going through the end of the boiler, returned to near the enlarged fire-place, when it passed out through the side of the boiler to the chimney, just as in the Tredegar puddling-mill boiler; all those variations were with the object of increasing the fire-grate, and at the same time keeping down the gross size and weight of boiler and its water.

[1] See vol. i., p. 104.   [2] See vol. i., p. 335.   [3] See vol. ii., p. 169.
[4] See vol. i., p. 181.   [5] See vol. i., p. 223.   [6] See vol. i., p. 207.

In 1805, Lord Melville failed to keep his appointment with Trevithick, on his proposal to construct a high-pressure steamboat.[1] Rennie, a pupil and friend of Watt, and familiar with Trevithick's high-pressure steam-dredgers on the Thames, was employed by Lord Melville and the Admiralty on the Plymouth Breakwater, where in 1813 Trevithick proposed the use of his high-pressure steam locomotive and boring engine.[2] In 1820 Rennie wrote to Watt, that the Admiralty had at last decided upon having a steamer; at that time fifteen years had passed since Trevithick's offer to propel the Admiralty by steam-puffers, and ten years more were to pass before they could make up their minds to venture on high-pressure steam from his boilers. The Steam Users' Association are equally hesitating, judging from words just spoken by an engineer, the son of an engineer :—

"Sir William Fairbairn said he had come to the conclusion, after many years' experience, that it was in their power to economize the present expenditure of fuel by a system which might not be altogether in accordance with the views of the members of the association or the public at large, and that was to increase the pressure of steam. He would have great pleasure in stating a few facts which might some day tend to bring about a change, if not a new era, in the use of steam. From the result of a series of experimental researches in which he had been engaged for several years on the density, force, and temperature of steam, he had become convinced that in case we were ever to attain a large economy of fuel in the use of steam, it must be at greatly-increased pressure, and at a rate of expansion greatly enlarged from what it was at present. Already steam users had effected a saving of one-half the coal consumed by raising the pressure from 7 lbs. and 10 lbs.—the

---

[1] See Trevithick's letter, 10th Jan., 1805, vol. i., p. 324.
[2] Ibid., vol. ii., p. 24.

pressure at which engines were worked forty years ago—to 50 lbs., or in some cases as high as 70 lbs. on the square inch."[1]

Dear me! would have been Trevithick's exclamation had he read this; did I devote my whole life to the making known the advantages of high-pressure steam, and did I, seventy years ago,[2] really work expansive steam of 145 lbs. on the inch in the presence of many of the leading engineers of the day! Of course this short extract of a speech made by a member of a practical society, may not be taken as conveying fully the speaker's views, but it illustrates the immense difficulty Trevithick encountered in making his numerous plans acceptable to the public.

Another modern statement bearing on inventions originating with Trevithick, but wearing new garbs with new names, shows the same tendency to ignore old friends, or, to say the least of it, to pass them by :—

"The trial of No. 36 steam-pinnace was made at Portsmouth yesterday. Her peculiarity consists in the arrangement of her propelling machinery, in the adaptation of the outside surface condenser, and a vertical boiler, both patented by Mr. Alexander Crichton. The condenser is simply a copper pipe passing out from the boat on one quarter at the garboard strake, and along the side of the keel, returning along the keel on the opposite side, and re-entering the boat on that quarter. The boiler is designed for boats fitted with condensing engines, and which, therefore, are without the acceleration of draught given by the exhausted steam being discharged into the funnel. It is of the vertical kind, and stands on a shallow square tank, which

_____

[1] 'The Engineer,' March 15th, 1872 : remarks by the Chairman at a meeting of the Manchester Steam Users' Association.
[2] See Trevithick's letter, August 20th, 1802, vol. i., p. 154.

forms the hot well. The tubes are horizontal over the fire, the water circulating through them. The condensed steam is pumped into the well at a temperature of 100°, and being there subjected to the heat radiating from the furnace, is pumped back into the boilers at nearly boiling point. It is estimated that, under these conditions, the pinnace would run for nearly 48 hours without having to 'blow off' or carry a supply of fresh water, the waste water being made good by sea water."[1]

The peculiarity of this steam-pinnace of 1871, on which a patent was granted, is stated to be a metal surface condenser exposed to the cold water at the bottom of the boat, returning the condensed steam at about boiling temperature to the boiler, and a vertical boiler with horizontal tubes through which the water circulates, both of which in principle, if not in detail, are seen in the surface condenser of Trevithick's iron-bottom ship of 1809, and his vertical boiler of 1816,[2] and further illustrated in the inventions spoken of in this and the following chapter; and yet on so all-important a subject, dealt with in various ways by Trevithick from 1804 to 1832, his plans are reproduced as discoveries in 1871.

About 1828, Mr. Rennie, Mr. Henwood, and others, reported on the advantages of high-pressure expansive steam in Wheal Towan engine,[3] on the north cliffs of Cornwall, near Wheal Seal-hole Mine on St. Agnes Head, where in 1797 Trevithick had worked his first high-pressure steam-puffer engine in competition with the Watt low-pressure steam-vacuum engine. Captain Andrew Vivian was then his companion, and the Cow and Calf, two rocks of unequal size, a mile from the land, were from that time called Captain Dick and

---

[1] 'The Times,' November 24th, 1871.
[2] See vol. i., pp. 336, 364, 370.
    See Mr. Henwood's report, vol. ii., p. 185.

Captain Andrew, or the Man and his Man, and there they still remain in the Atlantic waves, fit emblems of their namesakes and their still living inventions. The

CAPTAIN DICK AND CAPTAIN ANDREW, OR THE MAN AND HIS MAN. [W. J. Welch.]

stir made by those expansive trials led to the experiment in the 'Echo,' of which Mr. Henwood[1] thus speaks :—

"Captain William King, R.N., Superintendent of the Packet Station at Falmouth, attempted to impress on Viscount Melville, then First Lord of the Admiralty, the advantage of using high-pressure steam expansively in the Royal Navy, to whom Lord Melville replied that he had been taught by his friend, the late Mr. Rennie, that the danger attending such a course was very great, and that it would be difficult, if not impossible, to persuade him to the contrary."

Twenty-five years of precept and example caused the Admiralty to follow suit, and to request Mr. Ward, a Cornish engineer, to construct boilers and expansive valves for the Government steamboat 'Echo.' The writer was entrusted with fixing the machinery in the

---

[1] Residing at Penzance, 1871.

vessel at the Plymouth Dockyard, and before starting
with it from Harvey and Co.'s foundry, waited on
Captain King, R.N., at Falmouth, for his instructions,
in happy ignorance of the fear of the Lords of the
Admiralty to tread on Cornish high-pressure.   After
eying the applicant as captains in Her Majesty's service are apt to do when dealing with boys in the civil
service, he vouchsafed to say, " Mind, young man, what
you are about, for if there is a blow up, by —— you'll
swing at the yard-arm."

CHAPTER XXVII.

"Mr. Gilbert,
                                        "Lauderdale House, Highgate,
                                             "*March 1st*, 1830.

"Sir,—I have to apologize for my neglect in not calling on you, but ill-health prevented it. I left home on the 11th February, arrived in town on the 14th, and remained there until the 24th, when I was compelled to leave for this place, having a free good air. I am now taking, twice a day, the flowers of zinc, from which I hope to be soon right again. I am much better, but afraid to enter the city. I hope to be able to call on you before the end of this week, being very anxious to see you, having a great deal to communicate respecting the experiments I have been making, which will bear out to the full our expectations.

"Your hot-house apparatus has been finished nearly three months, all but two or three days' work to fit the parts together; I expect that before this they are in Penzance, waiting a ship for London. While making a sketch of your work for the founder, a thought struck me that rooms might be better heated by hot water than by either steam or fire, and I send to you my thoughts on it, with a sketch for your consideration. I find that steam-pipes applied to heat cotton factories, with 1 surface foot of steam-pipe, heat 200 cubic feet of space to 60 degrees. I also found in Germany, where all the rooms are heated by cast-iron pipes about the heat of steam, that 1 foot of external flue heated 160 cubic feet of space to 70 degrees.

"I find also that about 200 surface feet of steam-engine cylinder-case will condense about as much steam as will produce 15 gallons of water per hour, and will consume about 4 bushels in twenty-four hours to keep the temperature of

212 degrees. One bushel of coal will raise the temperature of 3600 lbs. of water from 40 to 212 degrees.

"A boiler, as the drawing, will contain 1200 lbs. of water, and consume one-third of a bushel of coal to raise the water from 40 to 212 degrees. It has 40 surface feet of hot sides giving out its heat. The 12-inch fire-tube in the boiler would raise the temperature to 212 degrees in about forty minutes. By these proofs it appears that 50 feet of surface steam sides will require 1 bushel of coal every twenty-four hours to keep up the boiling heat; therefore this boiler, having 40 surface feet, would give out the heat from one-third of a bushel of coal in twelve hours.

"Now suppose this charge of heat required to be thrown off in either more or less than twelve hours, the circular curtain would adjust the heat and time for extracting it.

"By the foregoing this coal and surface sides would heat to 60 degrees for twelve hours a space of 6800 cubic feet, equal to a room of 25 feet square and 11 feet high. If this boiler was placed in a room with a chimney, its water could be heated by having a small shifting wrought-iron chimney-tube of 4 inches diameter and 2 or 3 feet long attached to the end of the boiler while it was getting up steam, after which it might be removed, and the doors at both ends of the boiler closed; and as the boiler contains and retains its heat for twelve hours, more or less, it might be run on its wheels to any fire-place or

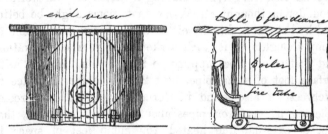

HOT-WATER ROOM-WARMER.

chimney to get charged with heat, and then run into any room, where there was no chimney, or into bed-rooms, offices, or public buildings; it would be free from risk, not having either steam or loose fire. The circular curtain, being fast to a wood

table, would by being drawn up or down adjust the required
heat and hide the boiler, and would be warm and comfortable
to sit at. I think this plan would save three-quarters of the
coal at present consumed; the expense of the boiler will not
exceed 5*l*. When you have taken it into consideration, please
to write me your opinion.

" I remain, Sir,

" Your very humble servant,

" RD. TREVITHICK.

"P.S.—Boiler, 3 feet diameter, 3 feet long; fire-tube, 12 inches
diameter, placed in the boiler, the same as my old boilers, made
of iron plates ⅛th of an inch thick, weighing about 2 cwt.

"I had a summons to attend at Guildhall last Saturday on
the coal trade, and was requested to attend a committee at
Westminster for the same purpose, in consequence of my applying small engines to discharge ships.

"I attended, but with difficulty, from my ill-health."

Trevithick was not above scheming for his friend's
hot-house, warming it by a boiler on wheels, in form
like his high-pressure steam-boiler. Rooms had before
been heated by steam or hot air in pipes; but he
thought a more simple and economical plan was to
heat a certain quantity of water to boiling heat at any
convenient place having a chimney, or in the open air,
and then wheel the apparatus into the room to be
warmed. If the room had a chimney, the fire could be
kept up, or the temporary iron connecting chimney
be removed and the apparatus wheeled into the middle
of the room and used as a table.

The scheme promised to be successful, for in a letter
nine months after the former he wrote that he had
taken a patent for France, where it had made a great
bustle among the scientific class, for coal in Paris was
3*s*. a hundredweight; some hot-water room-heaters

were the following day to be forwarded from London to
Paris; while the numerous orders were more than he
could execute. One in use at the 'George and Vulture'
Tavern, of a Gothic shape, handsomely ornamented
with brass, about two-thirds the size of the one in
Mr. Gilbert's hot-house, burns 7 lbs. of coal a day,
keeping the room at 65 degrees of heat during fifteen
hours. The rage amongst the ladies was to have them
handsomely ornamented.

Believing that they would be remunerative, he applied for the following English patent in February,
1831.

Apparatus for Heating Apartments. 21st *February,* 1831.

"Now KNOW YE, that in compliance with the said proviso, I,
the said Richard Trevithick, do hereby declare that the nature
of my said invention of a method or apparatus for heating
apartments, and the manner in which the same is to be carried
into effect, is shown by the following drawings and description,
where Fig. 1, Plate XVI., represents a longitudinal vertical section through the middle of a metallic vessel capable of containing
a considerable quantity of water, with a fire-place in the inside,
surrounded with water in all parts except at the doorway and at
an opening where the smoke may pass off into a common chimney.
Fig. 2, a vertical section near the fire-door, at right angles to the
section shown at Fig. 1; with the sections are also shown wheels
and handles, which lie out of the planes of the sections. The
letters of reference indicate the same parts in both figures. *a*,
the vessel; *b*, the space for containing the water; *c*, the fire-
place; *d*, the fire-bars, or grating; *e*, the ash-pit; *f*, an inner
door, to prevent the air from entering over the fire, yet allow it
to pass into the ash-pit, and thence up to the fire through the
grating; *g*, an outer door, to be shut when the fire is to be
extinguished; *h*, a chimney or flue, to convey the smoke into
a common chimney: this flue may be removed when the water
boils, and then the opening of the flue may be shut, to keep in
the heat, either by a door or by a plug fitting the opening;

PLATE 16.

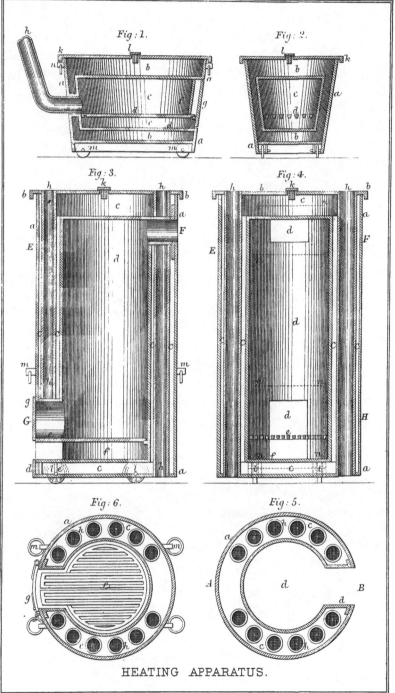

Fig: 1.

Fig: 2.

Fig: 3.

Fig: 4.

Fig: 6.

Fig: 5.

HEATING APPARATUS.

London: E. & F. N. Spon, 48, Charing Cross.   Kell. Bros. Lith. London

k, the cover of the vessel, having a rim all round, within which iron cement is to be driven to make the vessel steam-tight; l, a hole in the middle of the cover, into which a plug is dropped having a fluted stem and a flat head ground steam-tight upon the cover; this plug or valve is for the purpose of allowing the escape of steam if it should be raised above boiling point, and the valve is taken out when it may be necessary to pour water into the vessel; m, four wheels, on which the vessel may be easily removed from one room to another; n, two handles, to facilitate the removal. To use this apparatus for the warming of an apartment, the vessel is nearly filled with water, and placed so near to a chimney in another room, if more convenient, that the flue-piece h may convey away the smoke; a fire is then lighted upon the grating d, and continued till the water boils, when the flue-piece is taken away, and the flue opening stopped with the plug or door, and also the outer fire-door closed. In this state the apparatus is drawn into the apartment to be warmed, where it will continue for many hours to give off a most agreeable heat without any of that offensive odour usually experienced from stoves heated by an enclosed fire. Figs. 3, 4, 5, and 6 represent another form of my apparatus for heating churches or other large buildings. Fig. 3, a vertical section, from A to B, of Figs. 5 and 6, with a representation of the flue and its flanch, which lie beyond that section and the fire door-way and its flanch, which lie nearer, and also the four wheels, two of which are on each side of the section. Fig. 5, a horizontal section, from E to F, of Figs. 3 and 4. Fig. 6, a horizontal section, from G to H, of Figs. 3 and 4, with a view of the four handles situated at a higher level than the section, and of the fire-bars at a lower level; the same letters of reference signify the same parts in all the four figures. a, the outer case of the water-vessel; b, the cover; c, the space for water; d, the fire-place and flue; e, the fire-bars, made in two pieces, to be introduced through the fire doorway; f, the ash-pit; g, the fire-door; h, pipes open at top and bottom, cemented into holes in the bottom, and in the cover of the water-vessel; these pipes are to admit a current of air up through them, in order the more speedily to carry the heat into the building; k, the aperture in

the cover, to supply the vessel with water, and the plug to keep
in the steam; *l*, four wheels, on which the whole is moved, each
wheel revolving in a recess cast in the bottom of the outer case,
as represented by dotted lines in Figs. 3 and 4; *m*, four handles;
n, the flanches of the fire doorway and of the flue, represented in
Fig. 4 by dotted lines. A pipe to communicate with a chimney
while the water is being heated must be made to suit locality,
and therefore cannot require any description. This apparatus
can be heated in a vestry room, and the fire-door and flue closed
and then wheeled into the church, where it will soon diffuse a
most comfortable warmth; or the heat may be kept up while
standing in its place by having a constant communication with
a chimney, and thus diffuse a much more salubrious heat than
can be obtained by metallic or earthen stoves heated immediately
by the fire."

It is doubtful if the profits he received from the
heating apparatus covered the cost of the patent. The
first stove was not unlike his first locomotive boiler.
The more highly-finished stove resembled the marine
tubular boiler, also of former years, in the further
application of which we now follow him.

"Mr. GILBERT,   "HAYLE, *January 24th*, 1829.

        "Sir,—Since I have been down I have made a small
portable engine, and set it to work on board a coal-ship for
discharging the cargo; it is very manageable, and discharges
100 tons with 1 bushel of coal, without any person to attend it,
there being a string that the man in the hold draws when the
coal-basket is hooked, which is again drawn by the man who
lands the basket on the deck; the string turns and re-turns
the engine. It is near a ton weight, but as I find it double
the power required, I am now making a smaller one, $3\frac{1}{2}$ feet
high and $3\frac{1}{2}$ feet diameter, about 12 cwt.

    "I intend this engine to warp the ship, pump it, cook the
victuals, take in and out the cargo, and do all the hard work.
The captains are very anxious to get them on board every ship.
I think that an engine of 39 cwt. would propel their ships four

miles an hour over and above the other work of the ship, and
would neither be so heavy or take so much room as their
present cooking house and furnace. I think that two iron
paddles, one on each side of the rudder, under the stern, would
do this very well; they would be in dead water, and out of the
swell of the sea, and by being deep in the water would have a
good resistance. Two paddles, each about 4 feet deep and
3 feet wide, would do this, without their rising out of the water;
therefore their stroke would be nearly horizontal. The return
stroke would be in the water. Thus, let the paddle stand perpendicular in the water, two-fifths of its width on one side, and
three-fifths on the other side, the centre, which would turn
its edge to the water on its back stroke, and its flat to the water

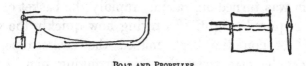

BOAT AND PROPELLER.

on the forward stroke; it would be light, and out of the way of
anything. I have a patent now going through the office for all
this, which will also cover the new principle of returning the
heat back again, as already described to you. The engine for
drawing in Holland will be ready about the end of February,
and by that time I shall have a complete portable engine ready
for London for discharging, when I shall be in town.

"I remain, Sir,

"Your very humble servant,

"RD. TREVITHICK.

"P.S.—Wheal Towan engine is working with three boilers,
all of the same size, and the strong steam from the boilers to
the cylinder-case; the boilers are so low as to admit the condensed water to run back from the case again into the boiler.
They find that this water is sufficient to feed one of these boilers
without any other feed-water; therefore one-third of the steam
generated must be condensed by the cold sides of the cylinder-
case, and this agrees with the experiments I sent to you from
Binner Downs. Wheal Towan engine has an 80-inch cylinder,

and requires 72 bushels of coal in twenty-four hours; therefore the cylinder-case must in condensing high-pressure steam use 24 bushels of coal in twenty-four hours. Boulton and Watt's case for a 63-inch cylinder, working with low-pressure steam, condensed only 4½ bushels of coal in equal time, the proportions of surface being as 190 to 240 in Wheal Towan. Nearly five times the quantity was condensed of high steam than of low steam, proving that there is a theory yet unaccounted for."

Trevithick's portable high-pressure steam-puffer engine, when it discharged the first cargo of coal from a vessel at Hayle, was worked by the writer; it stood on the wharf near the ship, and on a signal from the hold, steam was turned on, raising rapidly the basket of coal the required height. In trying how quickly the work could be done the hook missed the basket-rope, and caught the man under the chin, swinging him high in the air, much to the engineman's discomfiture. Fortunately the suspended man had the good sense to lay hold of the rope above his head, and so supporting his weight, no great harm was done.

The object and the means were the revival of the nautical labourer of twenty years before.[1] The boiler was a wrought-iron barrel on its end, on small wheels, with internal fire-tube, in shape like the boiler of the recoil engine of 1815;[2] but less high in proportion to its diameter. The cylinder was let down into the top of the boiler, and like Newcomen's atmospheric engine had no cylinder cover. The piston-rod was a rack giving motion to a small pinion fixed on a shaft on the top of the boiler, and to a large grooved wheel, around which was wound the whip-rope from the vessel's hold; a brake-lever enabled the engineman either to stop or

---

[1] See chapters xiv. and xv.
[2] See Trevithick's letter, 7th May, 1815, vol. i., p. 364.

to reduce the speed. Four months prior to the date of this letter he had sent a written offer to the Common Council of the city of London, offering to provide engines to discharge all coal-ships for the saving he would effect in six months, or he would supply an engine and boxes complete for 100 guineas. He at the same time suggested that in place of the baskets holding 1 bushel, iron boxes on wheels, holding 4 bushels, with a spring steelyard attached, should be used with his steam-engine, giving the exact weight without delay. He seems to have forgotten his nautical labourer patented twenty years before;[1] but yet reproduced something very similar.

Every trading vessel was recommended to carry at least a 12-cwt. high-pressure steam-puffer engine, suitable for warping, pumping, and discharging cargo; but a 30-cwt. engine, not occupying more room than a caboose, would in addition cook for the crew, and propel the vessel at three or four miles an hour.

Two iron paddles, like the duck's feet described to his Binner Downs friends many years before,[2] were to be fixed on an iron shaft across the stern of the vessel, receiving from the engine a motion like a pendulum. Each duck's foot was an iron plate 4 feet deep and 3 feet wide, turning partly round on its iron leg, to which it was attached as a vane, about 1 foot of its width on one side of its leg, and 2 feet on the other side; when the leg and foot were drawn toward the vessel, the foot, turning on its leg as a centre, exposed its edge only to the water; on the reverse movement, the longer side like a vane turned round until its flat was opposed to the water, in which position it was kept

---

[1] See vol. i., p. 325, and patent, 1809, vol. i., p. 302.
[2] See Mr. Newton's letter, vol. i., p. 342.

by a catch until the return movement, so that when it propelled, its whole surface pressed against the water, and when moving in a contrary sense, only its edge offered resistance to the water.

The writer has no record of the practical application of the duck's foot as a steamboat propeller; but the portable puffer-engine now pulls on board the fisherman's heavy nets, and the magnificent steamer 'Adriatic' hoists her sails on iron yards and masts by six of those steam helps.[1]

Twenty years before he had solicited the Navy Board to try his iron ships propelled by high-pressure steam-engines, and had shown their applicability as steam-dredgers; and again, shortly after his return from America, he pressed on their attention the same subject under new forms, followed by communications with their engineer, Mr. Rennie, and a proposal to place an engine in a boat at his own cost.

The writer has attempted in this and the preceding chapter to classify Trevithick's schemes, crowded together in those last years of his life, but the subjects so run into one another that the acts of twenty years before must be borne in mind to enable the more modern plans to be understood.

The letter introducing the surface condenser, in 1828, at the commencement of the former chapter, was in a month followed by that recommending a particular kind of paddle to be used as auxiliary steam-power, and after six months of experiments, by the patent of 1831, and the following correspondence :—

" MR. GILBERT,   "LAUDERDALE HOUSE, HIGHGATE, *June 10th*, 1830.

"Sir,—Yesterday I saw Mr. George Rennie, and he requested me to write to the Admiralty, a copy of which I send

[1] See 'Illustrated News,' 27th April, 1872.

both to you and to him, for your inspection. Mr. Rennie said there was a great deal contained in what I had stated to him, and that he would with pleasure forward my views, as far as he could with consistency.

"I remain, Sir,

"Your very humble servant,

"RICHARD TREVITHICK."

"To THE RIGHT HONOURABLE THE LORDS COMMISSIONERS OF THE ADMIRALTY, &c., &c., &c.

"MY LORDS,

"About one year since I had the honour of attending your honourable Board, with proposed plans for the improvement of steam navigation; and as you expressed a wish to see it accomplished, I immediately made an engine of considerable power, for the express purpose of proving by practice what I then advanced in theory. The result has fully answered my expectations; therefore I now make the following propositions to your honourable Board, that this entirely new principle and new mode may be fully demonstrated, on a sufficient scale for the use of the public.

"I humbly request that your Lordships will grant me the loan of a vessel of about two or three hundred tons burthen, in which I will fix, at my own expense and risk, an engine of suitable power to propel the same at the speed required. No alteration in the vessel will be necessary, and the whole apparatus required to receive its propelling force from the water can be removed and again replaced with the same facility as the sails, thus leaving the ship without any apparatus beyond its sides when propelled by wind alone, and when propelled by steam alone the apparatus outside the ship will receive scarcely any shock from the sea.

"This new invention entirely removes the great objection of feeding the boiler with salt and foul water, and not one-sixth part of the room for fuel, or of weight of machinery now used, will be required; it is also much more simple and safe, not only for navigation, but for all other purposes where locomotive power is required, and will supersede all animal power, as the objections of weight, room, and difficulty of getting and of carrying water in locomotive engines is entirely removed. It will therefore prove an investigation of greater utility to the public than anything yet introduced.

"I have to beg the great favour of your Lordships appointing not only scientific but practical engineers to inspect my plans, that you may be perfectly satisfied of their utility, not only in theory, but also as to the practicability of carrying the same into full effect."

The petition in June, 1830, for the loan of a Government hulk, hung fire up to January 1832, when an attempt was made to move the Lords Commissioners of the Admiralty by the force of numbers.

"We, whose names are hereunto subscribed, have known Mr. Richard Trevithick, of Hale, in the county of Cornwall, for a period of years, and during which time his conduct has merited our unqualified approbation. As an engineer of experience and eminence few, if any, can surpass him, and his present improvement of the steam-engine seems to outvie all others. We therefore, in justice to his talent, strongly recommend to the Lords Commissioners of the Admiralty that he may be permitted, at his own costs and charges, to fit and make trial of his engine in one of His Majesty's vessels.

" Dated in London this 27th day of January, 1832."

This was sent to Mr. Davies Gilbert, who on the same date suggested the following :—

" RECOMMENDATION OF MR. RD. TREVITHICK, January 27, 1832.

" We have not any doubt or hesitation in recommending Mr. Richard Trevithick as a man of extraordinary powers of mind, and of fertility of invention.

" Cornwall owes to him much of the improvements that have been made on Mr. Watt's engine—improvements that have reduced the consumption of coal to a third; nor have his exertions been confined to steam-engines alone. He now proposes to make the same water act over and over again by alternate expansion and contraction, which plan, if it succeeds, will be found of immense importance to vessels and locomotive engines.

" Understanding that Mr. Trevithick is desirous of making the experiment at his own expense, we clearly recommend that facilities may be afforded him." [1]

This paltry question with the Admiralty indirectly produced more trustworthy evidence of the great importance of Trevithick's inventions than all that has been written of him under the professional terms Engineers, and Engineering.

The names are not given of those who believed that he had, as an established fact, reduced the consumption of coal in the Watt engine to one-third; they were not Cornishmen, or they would not have misspelt the word Hayle, but they understood the great value of using the same fresh water over and over again in marine steam-engines.

Mr. Mills, who had taken an active part in the screw-propeller experiments in 1815, was again interested in the proposed trial in a Government ship, and wrote, "I have just left Captain Johnstone; he has communicated with Faucett and Co., Barnes and Miller, and with the firm of Maudslay. He has had his mind disturbed again by Maudslay about the greater quantity of water required to condense steam at higher temperatures; I repeated the same as yourself, about the cylinder full of steam, atmosphere strong; however, he appears quite different to what he was on Friday." Such a clique of professional friends would sink a stronger man than Trevithick. A year or two from that time the writer designed a high-pressure steam-engine suitable for a steamboat, and on presenting it to

---

[1] In the handwriting of Mr. Davies Gilbert.

the eminent marine-engine builders whom he served, was told that the lightness of the engine would cause less profit to the makers. Their bills were based on the pounds weight delivered, and new designs necessitated new patterns and new troubles. It was unreasonable to expect those makers of marine steam-engines to report that Trevithick knew better than they did. They knew of his screw-propeller experiments fifteen years before, but they in no way benefited him, and the Admiralty Captain was either a tool in their hands, or powerless without them.

The primary object, when the loan of the ship was asked, was the using for marine purposes a high-pressure steam tubular boiler, combined with tubular condenser, supplying or returning its water as feed, thereby avoiding the use of salt water in the boiler; and this steam-engine, as shown in his patent of 1831, was to be applied either to his screw, or his duck's foot, or other propeller; but during the year or two of suspense, other schemes for propelling ships had occupied his thoughts, resulting in the patent of 1832.

Steam-Engines, 1832.

" Now KNOW YE, that in compliance with the said proviso, I, the said Richard Trevithick, do hereby declare the nature of my said invention, as regards the improvement or improvements on the steam-engine, to consist in interposing between the boiler and the working cylinder, in a situation to be strongly heated, a long pipe formed of a compact series of curved or bent pipes, which I denominate the dry pipes, or steam-expanding apparatus, through which dry pipes I cause the steam, after it has been generated in the boiler in contact and consequently saturated with water, to pass with very great velocity, in order that it may imbibe a copious supply of additional heat without any addition of water, and by this additional heat to be expanded

into a greater bulk of steam, of about the same expansive force
that it had acquired in the boiler, by which means I obtain a
greater volume of steam for use in the working cylinder than
the boiler alone would supply; and in order still further to
augment this volume of steam, I place the working cylinder
within a case constituting a part of the flue or chimney, that
the cylinder may be kept considerably hotter than the steam
employed in it by absorbing a great portion of the heat remaining in the flue after having heated the boiler and the dry
pipes, which heat would otherwise pass away out of the top of
the chimney and be wasted, but by this arrangement is converted into a useful power by further expanding the steam in
the cylinder.

" And I do further declare, that in carrying this part of my
said improvement into effect, I do not find it necessary to confine myself to any particular form of boiler, or arrangement of
pipes, in which the steam is to be heated; but by preference,
as being very compact in form, and economical of fuel in using,
I make my boiler of a number of upright pipes, standing upon
and communicating with a tubular ring placed around and a
little below the fire-grate; these pipes all surround the fire-
place, except two or three, the lower ends of which are elevated
above the fire-door, but connected at the bottom by a branch
pipe united to one of the adjoining upright pipes, thereby
leaving an opening or place of access to the fire. These pipes
all extend upwards to the height of several feet, according to
the quantity of steam required to be raised, combined with
local convenience, for it is obvious that the power of this boiler
to raise steam may be increased either by increase of the length
of the pipes, of their diameters, or of their numbers. And I do
lay upon the upper ends of the pipes hereinbefore described
and connect with them a tubular ring similar to that upon
which the pipes stand, the two rings and the upright pipes
forming together a vessel in which water has free communication by means of the bottom ring to stand at the same level
in all the pipes, and the steam has free communication to pass
from all the pipes into the upper ring; and I do, for the sake
of obtaining great heat, place my system of dry pipes over the

fire, and within the circular row of upright pipes of the boiler hereinbefore described; and I form my dry pipes in pairs, each pair constituting the figure that is well understood by the term inverted syphon; and I unite several of these syphons together by short bent pipes at the top, so as to constitute one long zig-zag pipe, through which the steam must successively pass down and up the alternate legs of each syphon with great velocity, necessary for the rapid absorption of heat in its passage from the boiler to the working cylinder of the engine, the working cock, valves, or slide of which being united by a pipe of com-munication with that leg which is last in the succession of syphons; and I unite the first in succession of these inverted syphons with the upper tubular ring of the boiler by means of a bent pipe, in which a throttle-valve or cock is placed in order to limit the supply of steam, that it may have space in the dry pipes and working cylinder to expand in proportion as it receives additional heat; and I fix a safety-valve in communication with the boiler, and another in communication with the dry pipes; and I place around outside the boiler, at a small distance from the upright pipes, two cylindrical casings, one within the other, and fill up the space between the two casings with sand, ashes, or other material which conducts heat but slowly; and I close up the upper end of the casings over the boiler and the dry pipes with a covering in the form of a dome, and out of this enclosure I make the flue to pass to and around the working cylinder of the engine, whence the flue carries the smoke and little remaining heat away in any convenient manner; and I make my boiler-pipes, rings, and casings by preference of iron or copper, and my dry pipes of copper or other strong metal not liable to rapid oxidation by heat when in contact with steam; and I supply my boiler with water by means of a forcing pump, so adjusted as to keep the water of the proper height.

"And I do hereby further declare, that the nature of my said invention, as regards the improvements in the application of steam-power to navigation, consists in the drawing of water into a receptacle placed near within the stern of the navigable vessel, which water is drawn in through an orifice in the stern with a moderate degree of velocity in the direction of the course of the vessel, and ejected with great force and speed in a direction opposite to the course of the vessel through the same orifice, reduced to about a quarter of the area by means of a valve opening as the water enters, and partially shutting as the water is ejected; and thus I propel the vessel with great force, derived from the recoil of the water set into rapid motion in a direction opposite to the course of the vessel, the rapidity of the jet of water to be at least equal to double the required speed of the vessel to be navigated.

"And I further declare, that by preference I effect the purpose of receiving and of ejecting the water, and of deriving a motive force from its recoil, by means of a large vertical cylinder of cast iron or other metal, closed at both ends, in which a piston is forced up and down by a piston-rod sliding through a stuffing box in the lid, which piston-rod receives its motive force from a steam-engine; and I fix a tube into the after side of this cylinder, near the bottom, in communication with the space below the piston, which tube leads through the stern of the vessel, as low down as practicable, and opens on one side of the rudder; and I fix another tube into the after side of this cylinder, near the top, in communication with the space above the piston, which tube also leads through the stern of the vessel, as low down as practicable, but opens out on the other side of the rudder; and I place in the mouth of each of these tubes a valve opening inwards, to allow the water free entrance, equal to the bore of the tube, and partially shutting when the water is ejected, so as to reduce the opening through the stern to about one-fourth of the area of the tube.

"And I do hereby further declare, that the nature of my said invention, as regards the improvement in the application of steam-power to locomotion, consists in the application of such a boiler, together with the expanding apparatus as aforesaid, to locomotive engines, whereby a diminished weight of boiler and quantity of water and fuel is obtained; and in further compliance with the said proviso, I, the said Richard Trevithick, do hereby describe the manner in which my said invention is to be performed, by the following description of its various parts in detail, reference being had to the drawing annexed, and to the figures and letters marked thereon, that is to say :—

"*Description of the Drawing.* [Plate XVII.]

"Figure 1 represents a series of vertical sections through the various essential parts of the boiler, the dry pipes, the steam-pipe, the working cylinder, the propelling cylinder, and the flue, together with sections and views of other minor parts, serving to show the connections of the essential ones. The places at which these sections are taken are shown in Figure 2 by the dotted line from A to B, from B to C, from C to D, and from E to F. Figure 2 represents a plan of Figure 1, with the top coverings of the boiler and working cylinder removed. Figure 3 shows the manner of uniting the shorter upright pipes over the fire doorway with one of the adjoining ones, so as to give free circulation of the water in all the pipes. Figure 4 represents three pairs of syphons, which in their places stand in a circular form, but in this Figure are shown as spread out into a plane, in order the better to explain their structure and join-ings. Similar small letters and numbers of reference are used to denote similar parts in all the Figures; *a*, the upright boiler-pipes, the upright and lower ends of which are contracted to leave room for bolt-heads and nuts, without throwing the pipes too far asunder; *b*, the tubular ring having a flanch projecting inwards and outwards at the upper side, perforated with aper-tures upon which the upright pipes are bolted, and another flanch at the bottom, projecting inwards, to bolt the ring down to the foundation plate; *c*, the foundation plate; *d*, the fire-grate; *e*, the fire doorway; *f*, the upper tubular ring, having a flanch at the bottom projecting inwards and outwards, and perforated with apertures corresponding with the tops of the upright pipes upon which the tubular ring lies, and to all which it is bolted; *g*, the level of the water in the boiler-pipes; *h*, the dry pipes formed like inverted syphons, so as to require no joining at the lower part near the fire; one leg of each of the two syphons shown in Figure 1 is in section, and broken near the bottom; an outside view of the other leg appears partly behind the section; *k*, the short bent pipes, each bolted to two

PLATE 17.

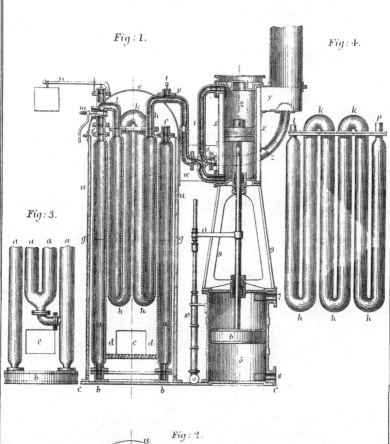

Fig: 1.

Fig: 4.

Fig: 3.

Fig: 2.

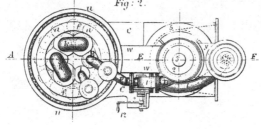

TREVITHICK'S PATENT BOILER AND ENGINE, 1832.

London E.& E N S₁m 46.Charing Cross.

Kell.Broˢ Lith. London.

syphons, to unite them into one continuous pipe; *l*, the bent pipe uniting the upper tubular ring with the first in succession of the syphons; the proper situation for this pipe is that shown in Figure 2, but for the sake of clearness and simplicity in the drawing, it is shown in Figure 1 as if on the left-hand pipe and syphon; *m*, the throttle-cock on the bent pipe *l*; *n*, the safety-valve lever, and weight on the same; *p*, the pipe of communica tion from the last in the succession of syphons to the working cylinder of the engine; *r*, the throttle-cock in the pipe *p*; *s*, a four-way cock, worked by the hand-gear, to direct the steam alternately under and over the piston; *t*, the safety-valve in communication with the dry pipes; *u*, the two cylindrical casings surrounding the boiler-pipes, the space between the two being filled up with a slow conducting medium; *v*, the domical covering over the cylindrical enclosure; *w*, the flue leading out of the enclosure into the casing of the working cylinder; *x*, the casing of the working cylinder forming a continuation of the flue; *y*, the further continuation of the flue to the chimney; *z*, the waste-steam pipe leading into the chimney; 1, the steam-pipes leading from the working cock into the top and bottom of the working cylinder; 2, the working cylinder; 3, the piston with metallic packing; 4, the piston-rod passing down through a stuffing box at the bottom of the working cylinder, and also continuing downwards, to form the rod of the propelling piston; 5, the propelling cylinder; 6, the water or propelling piston; 7, the upper aperture leading to one of the tubes opening through the stern of the navigable vessel; 8, the lower aperture leading to the other tube, opening also through the stern of the navigable vessel; these apertures are made as wide as the cylinder will allow, in order that they may have but little depth, and not occasion an inconvenient length of the propelling cylinder; 9, a frame supporting the steam-cylinder upon the propelling cylinder; 10, the feed-pump for supplying the boiler with water; 11, an arm fastened on the piston-rod to work the feed-pump and hand-gear; 12, the hand-gear.

"Now, whereas I claim as my invention, firstly, the interposing between the boiler and the working cylinder of the steam-engine a long many-curved heated pipe, through which the steam is forced to pass with great rapidity without being permitted to come in direct contact with water, by which arrangement the steam is made to absorb additional heat, and at the same time allowed to expand itself into a greater volume.

"Secondly, placing the working cylinder of the engine within such part of the flue or chimney as shall ensure the cylinder to be kept hotter than the steam used in it, by which means the expanding of the steam is still further promoted.

"Thirdly, propelling a navigable vessel by the force of the recoil produced from water received with a moderate degree of velocity, into a receptacle near within the stern, in the direction of the course of the vessel, and ejected with great velocity in a direction opposite to that course, the velocity of the jet being at least double the required speed of the vessel to be propelled, provided always that the same be effected in manner herein-before described. ·

"Fourthly, applying a boiler combined with a steam expand-ing apparatus, as before described, instead of a boiler alone, to a locomotive engine, whereby the power of the steam is applied after the steam has undergone the expanding process, and whereby a diminution is effected in the weight of the boiler, and in the weight and consumption of water and of fuel."

The two great objects in this 1832 patent were su-perheating steam in tubular boilers, and propelling ships by forcing a stream of water from the stern at a speed of at least double that of the vessel. Similar ideas may be traced in his patent of 1815, where a tubular boiler gave superheated steam, and in 1809 his patent for propelling steamboats "consists of a tube of considerable length disposed horizontally in the water, and the stroke of rowing is made by means of a piston with valves."

An engine of 100-horse power was ordered in Shrop-shire to be placed on board the Government ship to test the value of those patents of 1831 and 1832. One consequence was that a gentleman who had helped this scheme with his money wrote :—

"My case with Trevithick is strictly this; he was represented to me as a man of property; and as to his talents for mechanics, no man could be in his company long without being struck with them. I was induced to trust him to the amount of nearly 500l., and I then learned for the first time that it was only on the possible contingency of a grant from Government that he relied for the payment of my claim."

A company called the New Improved Patent Steam-Navigation Company was formed, of which Trevithick was a member, though apparently not a subscriber, for a note in November, 1831, informed him that "if in seven days he did not pay up his calls, his shares would be entirely forfeited." This company, among other proposals, opened negotiations for sending steamboats to Buenos Ayres to help in the commerce of the port and inland river.

In 1832 the Waterwitch Company made experiments with those plans, propelling by forcing water through pipes, since which a Government ship of war called the 'Waterwitch' has been so propelled. Twenty years ago the writer saw steamboats so propelled in daily use on the Meuse ; they needed no rudder, for by turning the mouth of the exit-water pipes on either side of the ship it was made to turn in its length, or even to move sideways.

Messrs. John Hall and Sons, of Dartford, also experimented on these two patents, and from this the tubular condenser was called Hall's Condenser. I think the boat it was first tried in was called the 'Dartford.'

Trevithick's difficulties in urging so many and great changes in marine propulsion may be estimated by the acts of other engineers.

"Mr. Rennie was engaged for many years in urging the introduction of steam-power in the Royal Navy. In 1817, we find
him writing to Lord Melville, Sir J. Yorke, Sir D. Milne, and
others on the subject. In July, 1818, he laments that he
cannot convince Sir G. Hope or Mr. Secretary Yorke of their
utility, but that he is persuaded their adoption *must* come at
last. On the 30th May, 1820, he writes James Watt, of Birmingham, informing him that the Admiralty had at last decided
upon having a steamboat, notwithstanding the strong resistance
of the Navy Board." [1]

So that Mr. Rennie, as professional adviser of the
Navy Board, had to persuade for three years, with a
knowledge of Trevithick's prior experiments, before
active steps were agreed to; for twelve years had then
passed since Trevithick's nautical labourer and iron
steamboat had been tried on the Thames, and five since
his experiments with the screw-propeller.

An article in 'The Times' gives in strong contrast
the relative value of screw and paddle-wheels as pro-
pellers. The 'Syria' was originally a paddle-wheel
steamer, having oscillating cylinders worked with steam
of 25 lbs. on the inch, and Hall's tubular condenser;
after a time the paddle-wheels were removed for a screw-
propeller, driven by two steam-cylinders side by side,
of different diameters, the high-pressure steam exerting
its full force in the small cylinder, and then expanding
in the larger cylinder. All the leading features in this
improved steamboat of the present day, such as high-
pressure expansive steam in one or two cylinders, with
tubular condenser and screw-propeller, had been publicly proved by Trevithick fifty years before.

"*Screw against Paddle.*—An interesting and important trial
trip has recently been made, which serves to exhibit the advan-

---

[1] 'Lives of the Engineers,' by Smiles, vol. ii., p. 267.

PLATE 18.

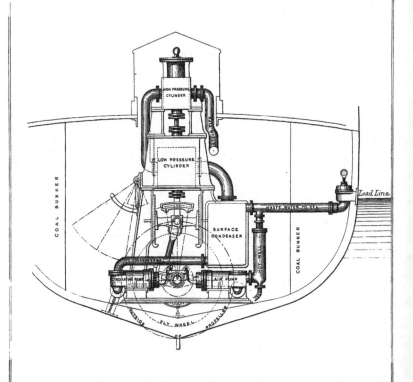

COAL BUNKER

HIGH PRESSURE CYLINDER

LOW PRESSURE CYLINDER

SURFACE CONDENSER

WASTE WATER TO SEA

Load Line.

COAL BUNKER

CIRCULATING PUMP

AIR PUMP

HOT WELL

OUTSIDE OF PROPELLER

FLY WHEEL

COMPOUND MARINE ENGINE, 1871.

London: E. & F. N. Spon, 48, Charing Cross.   Kell, Bro? Lith. London.

tages of the screw over the paddle as a means of propulsion for ocean-going steamships. In 1863, the steamship 'Syria,' of, 1998 tons, was built for the Peninsular and Oriental Company by Messrs. Day, Summers, and Co., and fitted with paddle-wheel engines of 450-horse power. The 'Syria' then attained a speed of 13·038 knots per hour, and the consumption of coal was at the rate of 45 tons per diem. The builders have lately converted her into a screw-steamer (for carrying the mails between Southampton and the Cape of Good Hope), who, without in any way disturbing the configuration of the hull, have fitted the 'Syria' with compound inverted engines of 300 nominal horse-power. These engines have two cylinders, respectively of 36 in. and 72 in. diameter, with a stroke of 4 ft. 2 in. On Monday last the 'Syria' attained an average speed of 12·637 knots, with a consumption of coal equivalent to 18 tons per diem; thus showing a difference of only 0·401 knot per hour, with a lessened power of 150 horses, and a saving in consumption of coal of 27 tons per diem; while the carrying capacity of the ship, arising from the economy of space in the engine-room, has been enormously increased, as she can now stow 1200 tons of cargo against 500 tons previously." [1]

Mr. Husband, of the firm of Harvey and Co., of Hayle, has obliged me with the annexed sketch (Plate XVIII.) of a modern high-pressure steam expansive compound marine engine, with surface condensers, on which the grandsons of Trevithick are now working, to be placed in the 'Batara Bayon Syree,' an iron yacht for an Indian Rajah, embracing the modern improvements of direct-action compound engines, and illustrating the principles which governed the constructors of the 'Syria.'

The first glance shows a seeming resemblance in outline to Trevithick's patent drawing of 1832, having one cylinder above the other; but a closer examination

---

[1] See 'The Times,' May 20th, 1871.

proves the application of the principles of his patents of
1815 and 1831, embracing screw-propeller, direct-acting
engines, tubular boilers, high-pressure steam used expansively, and condensation by cold surface preventing the
necessity of using salt water in the boilers.

This engine, in outline, has a strong likeness to
Trevithick's engines, going back even to his first patent
of 1802,[1] followed by the direct-action high-pressure
steam yacht of 1806,[2] and again in 1808[3] by the iron
steamer with direct-action long-stroke cylinders, with
highly expansive steam and surface condensers, to which,
in 1815,[4] was added the patent compound expansive
steam pole and piston engine and screw-propeller,
embodying during the first fifteen years of the present
century, both in principle and in detail, the most
approved form of marine steam-engine with fewness
and simplicity of form of moving parts; but compare
it with the Watt patent engine, and its difference
is obvious; no beam or parallel motion, no injection-
water necessitating the air-pump, no low-pressure
steam. The late Mr. William Wilson, of Perran
Foundry, son of Boulton and Watt's financial agent in
Cornwall, informed Mr. Henwood that he was with Mr.
Watt when some one stated that Mr. Trevithick was
working his engine with steam of 40 lbs. on the inch;
when Mr. Watt replied, " I could work my engine with
steam of 100 lbs. to the inch, but I [would not] be the
engineman."[5]

Progressive experience, with increasing demand for
economy and speed, have caused the principles and the
details of Trevithick's steam-engines to be matters of

---

[1] See vol. i., p. 59.   [2] See vol. i., p. 327.
[3] See vol. i., p. 336.   [4] See vol. ii., p. 103.
[5] Henwood, Address to the Royal Institution of Cornwall, 1871.

national importance seventy years after their discovery, for as far back as that he used highly-expansive steam,[1] and on the question of a separate cylinder for expansion as used in the modern steamboat combined engines, he wrote, " I think one cylinder partly filled with steam would do equally as well as two cylinders; that one at Worcester shuts off the steam at the first third of the stroke, and works very uniformly with a considerable saving of coal.[2] Those modern marine engines use about the same steam pressure and expand about in the same proportion. With the direct action from the piston-rod to the crank-shaft, the multitubular boiler and screw-propeller, and the surface condenser perfected in 1831 and 1832, at which time his construction of a marine steam-engine would have been just what it now is forty years later. Those latter patents also embrace the principle of superheating steam, practically shown many years before,[3] and still used by marine engineers of modern times.

In tracing the wisdom of his designs just before the close of an eventful life, reference may be made to the trial of a common road locomotive in 1871 :—" Experimental trip of the Indian Government steam train engine, 'Ranee,' from Ipswich to Edinburgh.— The results of the trial with the ' Chenah,' though satisfactory so far as the engines proper were concerned, were vitiated by the failure of the boiler; on the completion of the second engine, the ' Ranee,' the field boiler and variable blast-pipe were used; the boiler is about 4 feet diameter at the bottom and 8 feet high."[4]

The form and dimensions of the exterior of the Ranee

---

[1] See Trevithick's letter, 22nd August, 1802, vol. i., p. 153.
[2] See Trevithick's letter, 5th July, 1804, vol. ii., p. 132.
[3] See Trevithick's letter, 16th May, 1815, vol. i., p. 370.
[4] 'The Engineer,' 27th October, 1871.

tubular boiler are very similar to Trevithick's patent drawing and specification of 1832; even the variable blast-pipe was used by him in 1802.[1]

The last years of Trevithick's eventful life were chequered with hopes and disappointments when, in the early part of 1830, he wrote to his friend Gerard:—

"This morning I called here for the purpose of forwarding my information to the committee of the House. I called on Mr. Thompson to inform him what Mr. Gilbert said respecting it. His answer was, that the direct method would be by forwarding a petition in the way proposed when at the lobby. In consequence, I have forwarded the petition to Sir Matthew Ridley. Yesterday I took the coach to Highgate, by way of Camden Town, and of course had to walk up Highgate Hill. I found I was able to walk up that hill with as much ease and speed as any of my coach companions. However strange this maggot may appear in my chest and brain, it is no more than true. I wish among all you long-life-preserving doctors you could find out the cause of this defect, so as to remedy this troublesome companion of mine."

His health was breaking down, and his petition for a gift from the public purse, so hopefully commenced two years before, was doomed, after another year's bandying from pillar to post, to be forgotten and unanswered.

"DEAR TREVITHICK,   "EASTBOURNE, *December 26th*, 1831.

"I am sorry to find that you have not any prospect of assistance from Government. I have not any copy or memorandum of my letter to Mr. Spring Rice, but it was to the effect of first bearing testimony to the large share that you have had in almost all the improvements on Mr. Watt's engine, which have altogether about trebled its power; your having made a travelling engine twenty-eight years ago; of your having invented the iron tanks for carrying water on board ship. I

---

[1] Trevithick's letter, 22nd August, 1802, vol. i., p. 153.

then went on to state that the great defect in all steam-engines seemed to be the loss by condensation of all the heat rendered latent in the conversion of water into steam. That high-pressure engines owed their advantages mainly to a reduction of the relative importance of this latent heat. That I had long wished to see the plan of a differential engine tried, in which the temperatures, and consequently elasticities, of the fluid might be varied on the opposite sides of the piston without condensation; that the engine you had now constructed promised to effect that object, and that in the event of its succeeding at all, although it might not be applicable to the driving water out of mines, yet that for steam-vessels and for steam-carriages its obvious advantages would be of the greatest importance; and I ended by saying that although it was clearly impossible for me to ensure the success of any plan till it had been actually proved by experiment, yet judging theoretically, and also from the imperfect trial exhibited on the Thames, I thought it well worthy of being favoured.

"Your plan unquestionably must be to appoint some one with you, as Mr. Watt did Mr. Boulton, and I certainly think it a very fair speculation for any such person as Mr. Boulton to undertake.

"It is impossible for me to point out any individual, as never having had the slightest motive with such or with manufacturers in any part of my life, I am entirely unacquainted with mercantile concerns. I cannot, however, but conjecture that you should make a fair and full estimate of what would be the expense of making a decisive experiment on a scale sufficiently large to remove all doubt; and that your proposal should be, that anyone wishing to incur that expense should, in the event of success, be entitled to a certain share of your patent; on such conditions some one of property may perhaps be found who would undertake the risk, and if the experiment proved successful, he would be sure to use every exertion afterwards for his own sake. With every wish for your success,

"Believe me, yours very faithfully,

"DAVIES GILBERT."

The statement of the President of the Royal Society, that the power of the Watt engine had been trebled by Trevithick, brought him no gain.

He never troubled himself with politics, but the passing of the Reform Bill caused him to suggest that it should be commemorated by a pillar higher than had ever before been erected. The following memorandum is in his own writing :—

"' Morning Herald,' July 11th, 1832.

"*National Monument in honour of Reform.*—The great measure of Reform having become the law of the land, it is proposed to commemorate the event by the erection of a stupendous column, exceeding in dimensions Cleopatra's Needle, or Pompey's Pillar, and symbolical of the beauty, strength, and unaffected grandeur of the British Constitution.

" In furtherance of this great object, a public meeting is proposed to be held, of which due notice will be given, to set on foot a subscription throughout the United Kingdoms, limiting individual contributions to two guineas, but receiving the smallest sums in aid of the design.

"The following noblemen and gentlemen have signified their approbation of the measure:—His Grace the Duke of Norfolk, of Somerset, of Bedford; the Right Honourable Earl of Morley, of Shrewsbury, of Darlington; Lord Stafford; Sir Francis Burdett, M.P.; Joseph Hume, M.P.; R. H. Howard, M.P.; Wm. Brougham, M.P.; J. E. Denison, M.P.; A. W. Robarts, M.P.; J. Easthope, M.P.; General Palmer, M.P."

" Design and specification for erecting a gilded conical cast-iron monument. Scale, 40 feet to the inch of 1000 feet in height, 100 feet diameter at the base, and 12 feet diameter at the top; 2 inches thick, in 1500 pieces of 10 feet square, with an opening in the centre of each piece 6 feet diameter, also in each corner of 18 inches diameter, for the purpose of lessening the resistance of the wind, and lightening the structure; with flanges on every edge on their inside to screw them together;

seated on a circular stone foundation of 6 feet wide, with an
ornamental base column of 60 feet high; and a capital with
50 feet diameter platform, and figure
on the top of 40 feet high; with a cylinder of 10 feet diameter in the centre of
the cone, the whole height, for the accommodation of persons ascending to
the top. Each cast-iron square would
weigh about 3 tons, to be all screwed
together, with sheet lead between every
joint. The whole weight would be about
6000 tons. The proportions of this cone
to its height would be about the same
as the general shape of spires in England.

"A steam-engine of 20-horse power
is sufficient for lifting one square of iron
to the top in ten minutes, and as any
number of men might work at the
same time, screwing them together, one

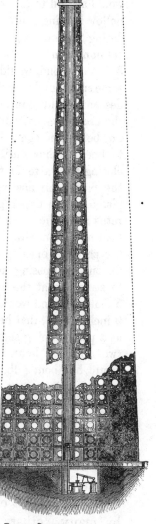

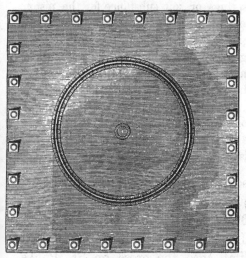

PLAN AND SECTIONAL ELEVATION OF PROPOSED REFORM COLUMN.

square could easily be fixed every hour; 1500 squares requiring
less than six months for the completion of the cone. A proposal

2 D 2

has been made by iron founders to deliver these castings on the spot at 7*l*. a ton; at this rate the whole expense of completing this national monument would not exceed 80,000*l*.

"By a cylinder of 10 feet diameter, through which the public would ascend to the top, bored and screwed together, in which a hollow floating sheet-iron piston, with a seat round it, accommodating 25 persons; a steam-engine forces air into the cylinder-column from a blast-cylinder of the same diameter and working 3 feet a second, would raise the floating piston to the top at the same speed, or five or six minutes ascending the whole height; the descent would require the same time. A door at the bottom of the ascending cylinder opens inwards, which, when shut, could not be opened again, having a pressure of 1500 lbs. of air tending to keep it shut until the piston descends to the bottom. By closing the valve in the piston it would ascend to the top with the passengers floating on air, the same as a regulating blast-piston, or the upper plank of a smith's bellows. The air apparatus from the engine should be of a proper size to admit the floating piston with the passengers to rise and fall gradually, by the partially opening or shutting of the valves in the top of the piston. Supposing no springs or soft substance for the piston to strike on at the bottom of the column-cylinder, descending 3 feet a second would give no greater shock than falling from 9 inches high, that being the rate of falling bodies, or the same as a person being suddenly stopped when walking at the rate of two miles an hour. The pressure of the air under the piston would be about $\frac{1}{2}$ lb. on the square inch; the aperture cannot let the piston move above 3 feet a second, but this speed may be reduced to any rate required by opening or shutting the valves on the floating piston."

To Trevithick's soaring genius nothing appeared very small, or very large, or very costly; not even the cast-iron column 1000 feet high covered with gold. The stone monument of London, 210 feet high, is admired by many; others climb into the cross on St. Paul's Cathedral, 420 feet high; some make a long journey to the great Pyramids, 500 feet high. How much

more pleasant would be Trevithick's proposed floating
1000 feet upward on an air-cushion, controlled by his

GENERAL VIEW OF REFORM COLUMN.

high-pressure steam-engine, and having, from the loftiest
pedestal of human art, surveyed imperial London, to be
again lowered to the every-day level at a safe speed,
regulated by valves closed by such simple acts as rising
from the seat; but should this be neglected, the passage

of compressed air escaping from under the piston-carriage would only allow of its descent at a speed of 3 feet in a second, giving but the same shock on bumping the bottom as jumping off a 9-inch door-step.

Perhaps the King in 1833 could not take an active part in advocating a memento of the golden days of reform; but this is no reason why the suggestion should have been so slightly noticed in 1862, to erect it in memory of the good and wise Prince Albert.

Various meetings were held, and after nine months the plan had so far advanced as to be placed before the King.

"Sir Herbert Taylor begs to acknowledge the receipt of Mr. R. Trevithick's letter, with the accompanying design for a national monument, which he has had the honour of submitting to the King.

"St. James's Palace, 1st *March*, 1833." [1]

Within two months from the date of the design for a gilded column Trevithick had passed away. His family in Cornwall received a note, dated 22nd April, 1833, from Mr. Rowley Potter, of Dartford, stating that Trevithick had died on the morning of that day, after a week's confinement to his bed. He was penniless, and without a relative by him in his last illness, and for the last offices of kindness was indebted to some who were losers by his schemes. The mechanics from the works of Messrs. Hall were the bearers and mourners at the funeral, and at their expense night watchers remained by the grave to prevent body-snatching, then frequent in that neighbourhood.

A few years after the funeral, the writer was refused

---

[1] The column was suggested in 1862 as a suitable monument to the memory of the late Prince Albert.

permission to go through the works to inquire into the character of the experiments that had been tried, but the working mechanics were glad to see the son of Trevithick, and their wives and children joined in the welcome as he passed through the small town.

Trevithick's grave was among those of the poor buried by the charitable; no stone or mark distinguished it from its neighbours. He is known by his works. His high-pressure steam-engine was the pioneer of locomotion and its wide-spreading civilization. England's mineral and mechanical wealth on land or sea are indebted to its expansive power, its applicability, and durable economy.

His comprehensive and ingenious designs, given to the world seventy years ago,[1] are still instructive guides; and many of his works, dating from the dawn of the present century, remained as active evidences of his skill almost to the present day, with their three-score years,[2] while some few reaching three-score years and ten still remain good servants[3] in the solitude of the Peruvian mountains, where no mechanical hand repairs the errors of human skill or the wear and tear of time.[4]

If these material proofs fail to convince, the reader has but to ponder on the bitterly natural reflections written by himself a few months before his last illness to his friend Davies Gilbert:—

"I have been branded with folly and madness for attempting what the world calls impossibilities, and even from the great engineer, the late Mr. James Watt, who said to an eminent scientific character still living,[5] that I deserved hanging for bringing into use the high-pressure engine. This so far has

[1] See 1802 patent, vol. i., p. 128.   [2] See vol. i., pp. 222, 100, 82, 184.
[3] Agricultural engine, vol. ii., p. 68.   [4] See vol. ii., p. 220.
[5] Mr. John Isaac Hawkins.

been my reward from the public; but should this be all, I shall
be satisfied by the great secret pleasure and laudable pride that
I feel in my own breast from having been the instrument of
bringing forward and maturing new principles and new arrangements of boundless value to my country. However much I
may be straitened in pecuniary circumstances, the great honour
of being a useful subject can never be taken from me, which to
me far exceeds riches."

INDEX TO VOLUME II.

THE END.

LONDON: PRINTED BY W. CLOWES AND SONS, STAMFORD STREET AND CHARING CROSS

Printed in the United States
By Bookmasters